# ZETETIC ASTRONOMY:

OR

## THE SUN'S MOTIONS NORTH AND SOUTH;

WITH

## THE MOON'S MOTIONS;

FANCIED AND REAL.

SHOWING THE USELESSNESS OF THE GRAVITATION THEORY,
etc.

## By Lady Blount and Albert Smith.

*Price, One Shilling ; by post, 1s. 1d.*

Printed and Published by LADY BLOUNT, II, Gloucester
Road, Kingston Hill, Surrey,
TO WHOM THE COPYRIGHT BELONGS.

# Table des matières

# PREFACE

It was the late Richard Proctor who defined astronomy as "a science whose facts are based upon reasoning." Inasmuch as the great temptation of all scientists is to conclude their reasonings before they have obtained all their data the result is that their facts are based on insufficient evidence.

Our object, in the following chapters, is to examine these "reasonings," and to show that their conclusions are false, and their facts non-existent. While, as Zetetics, we are not called upon to explain phenomena, or construct hypothetical astronomical systems, it is our duty to show how far popular hypotheses fail, because the popular theory is made the basis of attacks upon the cosmogony of the Bible. We submit the following chapters to our fellow Zetetics as an honest attempt to explain celestial phenomena, especially in the South, which have not yet been explained on Zetetic lines.

But while admitting known facts connected with celestial phenomena, we cannot see that one single fact in this direction has tended to throw any discredit on our basal fact: that the surface of all standing water is level, or horizontal; and thus we undeniably prove that the earth (i.e., the earth and sea together) is not a globe nor a luminous star in the heavens. The Scriptural order of Creation is set forth in the second of the Ten Commandments, viz .:

Heaven *above*, the
Earth *beneath*, and
Water *under* the Earth.

Christians should remember the words of their Lord (see John v. 46, 47,) and rest assured that the Bible is as scientifically accurate in its account of Creation, as it is in its Plan of Redemption by Jesus the Christ, which God, that cannot lie, promised before the world began."

# CHAPTER I: THE SUN'S MOTIONS NORTH AND SOUTH

Zetetics, who derive their name from *Zeteo*, to search out or to investigate, may fairly claim that they have frequently and practically proved that the surface shape of the earth and sea are, generally speaking, horizontal. Every copy of *The Earth* gives proofs of this fact. Then, when tangible proofs are given, objectors, instead of considering the evidences brought forward, go off into celestial phenomena. Even some whose education would lead us to suppose that they had, to some extent at least, cultivated the logical faculty, act in this manner. Thus we are to some extent driven to consider celestial phenomena with a view to meeting objections, or answering enquirers. Education, as conducted on modern lines, does not always conduce to the bringing out of the logical faculty. So by way of introduction we must emphasize the fact that if we can give only one proof that the earth is a motionless plane, no other fact in Nature can controvert or overthrow that primary fact; but the fresh fact must be explained, if explained at all, in harmony therewith.

Now Lady Blount's late photographic experiment on the Bedford Canal, with a Dallmeyer photographic lens, conducted by an expert photographer under the direction of her ladyship, has undoubtedly given Zetetics printed proof of their basal fact, namely that water is level, and the earth therefore a plane. This was a great service rendered to the truth, for which due credit should be given to her, both by Zetetics and Globularists. The experiments were conducted openly by an expert photographer at considerable expense of time and money to her ladyship, for no personal gain, but simply with the one object of illustrating the truth. This should show our opponents that we are sincere, whether they are so or not. It is hard to believe that some of our critics are sincere, for they make no effort and are at no expense nor trouble to find out the truth in this matter. But sitting perchance in an editorial chair, or maybe simply writing as private and irresponsible critics, they urge their weak and sometimes fallacious objections. For instance one editor of a photographic journal speculates as to what the account of the experiments may have arisen from, as though to suggest that he was not sure that the experiments were made ! He should acquaint himself with the subject before he writes upon it. Then, on the supposition that the experiments were performed, he proceeds to explain away the results, saying: "On the other hand, unusual or special atmospheric conditions of refraction often step in, and render objects visible which are considerably below the horizon." This is the old trick of mere partisans, who always hold stubbornly to their own views, whatever evidence is produced. Another, a private correspondent, who professes to be critical, though he is not always logical, ignoring the zeal, trouble, and expense of conducting the experiment, writes, coldly harping upon the same monotonous strain, "refraction." He confesses that as a globularist he was somewhat staggered by the conclusive evidence there obtained, until he was reminded by a letter in *The Earth* of some "mathematical tables," giving tables of "correction for refraction"! Though as the Ed. then very properly added in a foot-note, "proof should be first given that any correction was needed over a level surface, where the rays of light would travel through a medium of almost unvarying density." But though they have no proof that there was any correction needed, they seem to think that the possibility of such is enough without any evidence and so they sit still and cry out "Refraction" ! It is amusing. But when the ship disappears at sea, *that* is not caused by refraction but "curvature" ! But when the ship is shown through a good glass, or a signal close to the water's edge six miles away, they then again shout "Refraction" ! Thus, like the man in Aesop's Fables, they can blow both hot and cold. But we must leave dishonest critics to their delusions, and try as best we can to help true enquirers.

The questions most commonly asked of late, are such as the following. Has a midnight-sun been seen in the south? Is it reconcilable with the plane-earth teaching? Do degrees converge or diverge south of the equator? And what then must be the motion, or motions, of celestial bodies, and especially of the sun in southern latitudes? In the following articles we shall try to answer these questions according to the best light we have received up to the present, and of course in harmony with the ascertained fact that the earth is a plane. We must start with facts, and endeavour to make logical deductions from them ; and we must remember that we are dealing with celestial phenomena rather than with terrestrial.

## THE SUN'S MOTIONS NORTH

Let us start with the motions of the sun North, for it is with these that we are most familiar. On June 22nd, this year, 1904 A.D., the sun entered the tropical sign of Cancer. It then attained its furthest North declination, or distance from the celestial equator, 23° 27'. It also then attains its highest noon altitude in countries situated like England, and those still further north. Hence the northern summer then begins. But the sun only remains at this declination for a short time. It begins to enlarge its daily circuit round the northern portion of the earth. We will illustrate its motion by a diagram.

DIAGRAM I

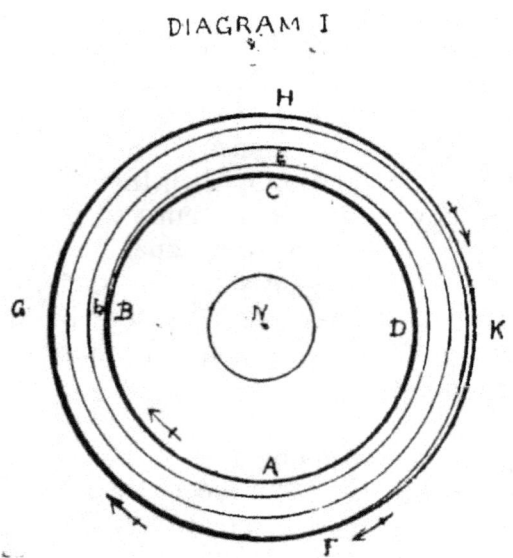

In the above diagram A B C D represents what is usually called the Tropic of Cancer.

It would be more correct to say it represents the path of the sun for that one day when the sun enters the first degree of the celestial sign Cancer. The sun moves round in its northern circuit in the direction of the arrows, that is, supposing it to start at A, it goes on to B in six hours, to C in twelve, to D in eighteen, and back again to A in twenty-four. But when the sun gets back to A it begins to leave the circle A B C D, and gradually recedes further from the centre N, which Zetetics call the North Centre.

In fact the circle A B C D is the only circle which the sun makes for six months, until it makes a similar circle in the South. The circle the sun makes in Cancer then begins to enlarge, and leaving the circle A B C D, the sun next courses from A towards (b) and on to E, &c., in a spiral movement which is almost circular but not quite so. Its declination varies one or two minutes per day to the end of June, and more rapidly afterwards, until the sun gets back to the equator F G H K, when of course it has little or no declination. Thus in three months the sun arrives at the equator, making in this time about eighty-nine daily revolutions round the northern parts of the earth. We have only shown three spiral lines in the diagram between the tropic of Cancer and the equator, because it would manifestly overcrowd the diagram to make eighty-eight or eightynine circles. But if we remember that the sun makes about thirty different revolutions per month, we shall see that it is a very fine spiral line which would be required to exhibit the sun's path for this period.

That the sun moves daily round us anyone can see from his own observation ; and though many tests have been applied by Zetetics, the earth has never been found to have any motion, that is, such as astronomers call its "diurnal motion." To discern that the sun's motion is *spiral* closer observation is required, with daily comparisons of its position when rising, culminating, and setting.

But even impartial globularists have confessed to this spiral-like movement of the sun, when, forgetting their globular theories, they honestly describe Nature as they really see her. For instance, in an interesting book by Paul B. du Chaillu, entitled *The Land of the*

*Midnight Sun,* he says: "The sun at midnight is always NORTH of the observer (fact) on account of the position of the earth" (theory).

"It seems to travel in a 'circle' (fact)....... At the pole the observer seems to be in the centre of a GRAND SPIRAL MOVEMENT OF THE SUN, which further south takes place north of him."

This agrees well with the plane truth, but it is out of harmony with the globular theory, as was shown many years ago by "Zetetes" in his pamphlet on *The Midnight Sun* (north).

## THE SUN'S MOTION SOUTH

We next proceed to give some evidence of the sun's motions in southern regions. Here we shall have to depend upon the evidence we have gathered for some time past, both from Zetetics in southern latitudes, and also from others who are globularists.

One correspondent in E. Australia, an intelligent Zetetic, and formerly a teacher says:

"When I stand with my face to the North, the sun rises in the south-east, and travels from my right hand to my left almost straight overhead but a little in front of my face, and then sets in the southwest. This is in the height of our summer—Christmas-time. The south side of buildings gets the sun in the mornings and evenings in summer, but not in winter, as the sun rises more north-east and sets more north-west; and it does not rise nearly so high overhead."—R.A.

This is good general testimony, and it agrees with other reliable, and perhaps more "scientific" testimony which we received from the Perth Government Astronomer in West Australia, some of which lately appeared in *The Earth*. It also agrees with evidence from a Zetetic, printed in *The Earth (not a globe) Review* so far back as 1893. That Zetetic, Mr. George Revell, further said:

"The Southern Cross and all other Constellations do most certainly appear to revolve around a southern point or centre. I have proved this beyond doubt by close observation....... the circle seems to narrow in winter, and expand in summer."

This is important testimony, and we quote it from Zetetics in the South because we believe it will appeal more forcibly to Zetetics in the North than would the testimony of those opposed to the plane truth.

These southern Zetetics know that the earth and sea are horizontal and stationary, yet they are candid enough to testify to celestial motions which some illogically think are opposed to this great fact. But one fact can never contradict another fact: both must be true. Zetetics therefore in the North must be candid enough to accept the facts on celestial motions in the South, just as we wish globularists to be candid, and reasonable enough to accept the wellknown fact that water is level, and the earth therefore a plane. Only those who are candid and sincere will arrive at all the truth; though they may not obtain it "all," they certainly will obtain much more than those who are not candid.

We shall (D.V.) give some further evidence respecting southern celestial phenomena in our next article, and attempt to illustrate the same by further diagrams. As we write chiefly for Zetetics we shall close this chapter with a quotation from one who, according to our Lord was inspired by the Spirit of God when he wrote:

"The heavens declare the glory of God, and the firmament showeth His handiwork. Day unto day uttereth speech; and night unto night showeth knowledge. There is no speech nor language where their voice is not heard. Their line (margin : *rule*) is gone out through all the earth; and their words to the end of the world. In them hath he set a tabernacle for the sun, which is as a bridegroom coming out of his chamber; and rejoiceth as a strong man to RUN A RACE. His going forth is from the end of the heaven, and his circuit unto the ends of it: and there is nothing hid from the heat thereof."—*Ps.* Xix. 1-6.

These wonderful words contain some valuable hints which we may further explain as we proceed with later chapters.

# CHAPTER II: THE SUN'S MOTIONS SOUTH

Already we have given evidence on this subject; and evidence from Zetetics which should appeal to Zetetics.

Let Zetetics weigh that evidence carefully for it will prepare them for that which follows. But we have further evidence from honest and skilful opponents. Every fact should be acknowledged by us whether it come from friend or foe. Reverence for facts should be a characteristic of Zetetic investigation. This being so we shall have to admit that the sun does *not* enlarge its circles south of the equator. (See the following quotations from a letter lately received from Auckland, New Zealand):

"It was because I found on my recent visit to England that some of my friends denied the existence of a southern centre that I had the photos taken for their benefit......The same circumpolar stars visible from my house, I have seen from Sydney,Melbourne,Adelaide, and Capetown, .......Looking north all the stars rise in the East (to the right) and set in west (to the left).......Yes. the length of the day increases in summer and shortens in winter as we go south."—G.A.

We have seen star photos taken by the writer of the above extract, with his camera turned towards the south centre, and the stars have made traces on the negatives and the photos which are evidently parts of circles ; thus proving that they are circumpolar. There are therefore circumpolar constellations south as well as north of the equator.

The following constellations may be mentioned as being entirely circumpolar: Octans (in which is situated the southern "pole" star, Sigma Octantis); also the whole of Mensa, Musca, Chameleon, and Triangulum. There are also portions of the southern constellations Argo, Crux, Centauras, Paro, Indus Tucana, which never set. These may be seen in their relative positions in any good star atlas which gives the southern constellations; for instance, *The Midnight Sky, looking south,* by E. Diinkin; though there may be later and better atlases.

Now the above facts prove that these constellations in the south move round a southern point near the small star, Sigma Octantis. We are credibly informed that "they all appear to move round this point." Zetetics will readily believe that they move, not the earth!

The fact that they so revolve proves the further fact that the ethereal currents which carry them round are similar to those currents which exist in the North, There are therefore two sets of such currents instead of one, as formerly supposed. But more of this anon. We shall now proceed to give a diagram illustrating these facts.

## DIAGRAM 11

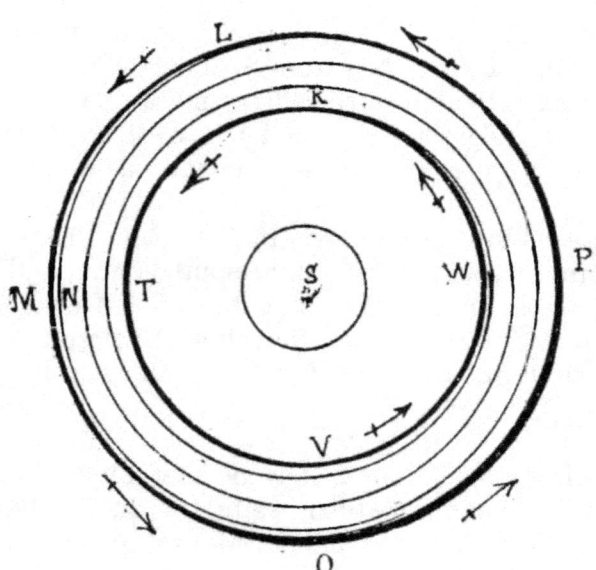

In the former diagram we traced the sun's spiral motion from the Tropic of Cancer to what is popularly known as "The Equator." It must be remembered, however, that the Equator is only an "imaginary line," generally given as a circle, which is supposed to represent the sun's path at the equinoxes. Let us therefore start our diagram so that the sun is said to be on or above the equator, say at L in diagram II. Later on we may connect this with diagram I. The sun still goes on with its "spiral motion," not from L to M, as is popularly supposed, but from L to N, on the inner curve about 23' (23 minutes) to the South. Then as the sun's declination South still increases, that is as the sun goes further south and therefore nearer to, the southern centre, its spiral movement carries it round that point until it arrives at R, on the inner circle. This is technically called the Tropic of Capricorn. The sun arrives at the first point of the southern sign, Capricorn, this year (1904) on December 22nd, about mid-day, Greenwich time; it then describes the circle R T V W in the direction indicated by the arrows. This circle and a similar one in the north are the only two circles the sun ever describes; all the rest of its revolutions, or nearly all, are spirals, very fine spirals certainly, but being so they are not exact circles.

Now, when the sun reaches the southern tropic, Capricorn, it is summer in the south, though at Christmas-time it is winter with us in the north. The sun's daily circuit is then at its nearest to the southern centre; and so the mid-night sun may be seen in the South, at this time of the year, just as it may be seen in the North during the Arctic Summer. We have already shown how this is possible over a planeearth, and how it conflicts with the idea that the earth is a sphere: (see articles on "The Midnight Sun, both North and South," which appeared in *The Earth*, vol. i., Nos. 4 & 5, 1900).

That the midnight sun was seen South was reported by the late Antarctic Expeditions. In fact it was reported that the sailors played cards on deck in full sunlight at midnight about Christmas-time (see reports in various newspapers, afterwards; see also *The Windsor Magazine* for May, 1900, in which a photograph— presumably— is given of the Midnight Sun, taken from the ship *Belgica*, Christmas, 1898.)

The *Belgica* was frozen fast in the ice from March 4th 1898, until February, 1899. The sun is represented as well above the horizon, so that it could shine "down" on the sailors upon deck at midnight. Though not in harmony with the globular theory, this may be taken as fairly representing the facts of the case, and as harmonizing with other known facts which have been reported to us both by Zetetics and Globularists. For instance we have made inquiries from various persons, and we find that the longest day in any place further south than the tropic of Capricorn increases in length as the latitude increases, or distance south of the observer. It follows therefore, logically, that if we were to go further and further south the longest day would keep on increasing until it filled the whole of the twenty-four hours. Then of course the Midnight Sun might be seen.

Testimony has now been given that it has been seen, and this testimony has been admitted in Zetetic literature, thus proving, speaking generally, that Zetetics are willing to learn and to admit of known facts.

Whether we can explain these facts is another question. But, as we have before intimated that whatever further facts we may find, which are proved to be real facts, and not fancies, we shall admit, but while admitting them to be true Zetetics will never give up their primary fact that the surface of all still water is level, and the earth, or land portion of the world, therefore a plane or series of planes.

We must now proceed, and follow the sun's course still further, that is back again to the Tropic of Cancer, from whence we started. To do this we shall have recourse to further diagrams.

DIAGRAM III

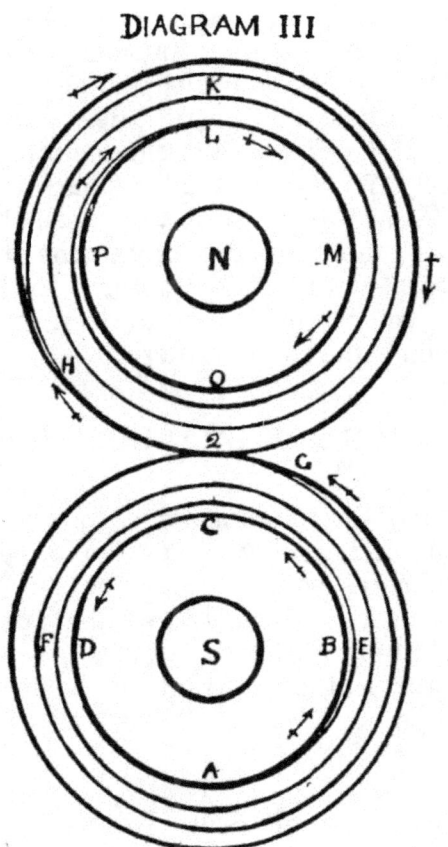

We have traced the sun's path for six months of the year, from the Tropic of Cancer to the Tropic of Capricorn. These are called "tropics" because the sun turns back towards its former course and position. The word "tropic" comes from the Greek work *trepo*, which means to turn. These turnings take place when the sun is 23 ½° from the equator, whether North or South, or about 66 ½° from the central points.

Taking the southern circle in Diagram III., we now proceed to consider the sun circling round the southern point, S. This takes place in the present year (1904) on the 22nd December; that is, the sun enters the cardinal sign Capricorn on this day, and its declination is practically the same for two or three days. Then it begins gradually to turn further away from the Southern Centre. For some days its declination alters very little— not more than about one minute of a degree per day. It circles round the southern point in the direction A B C D, as indicated by the arrows. It is like a great clock spring beginning slowly to unwind itself. Leaving the inner circle the sun may be represented as proceeding gradually from A to E, and so on unto F, until it finally returns to the equator at the point G. Here we will leave it for a moment to consider another important fact, namely the fact that there are

## TWO GREAT CURRENTS.

Whether North or South of the imaginary line called the "Equator," we have seen that the heavenly bodies, being light and small, are carried round their respective centres by the ethereal currents which prevail around these centres. We may speak of these more fully later on, and attempt to show what is the natural or physical cause of the direction of these currents, as secondary causes under the first great cause of all things, the fiat of the Ever-Living and Ever-Blessed Creator.

Our study of these currents will give us further reasons for disbelieving the commonly accepted "Theory of Gravitation," as being not only absurd in itself but utterly inadequate to produce results which are in evidence everywhere. But this we must omit for the present, our object now being to intimate partly how the sun passes from one current into another when, to use a popular expression, it "crosses the equator." These great currents, flowing around their respective centres, and outward therefrom, will, of course, meet and combine in the "equator " or their outer circumferences. Here, to some extent, they mutually check each other, and there is therefore what is known as the "region of calms" upon the earth's surface. But while the sun is in the south for instance, and coming towards the equator, its circle, or *circuit* rather, keeps on expanding, and thus the sun acquires a sort of momentum outwards. Now we left the sun, for a moment's consideration with this acquired momentum, at the point G. This momentum carries the sun from the point G onwards to meet the currents circling round the North Centre at Q. It is thus transferred from one set of currents to the other, and, "crossing the equator," the sun pursues its course northward towards H, and so enters its northern declination, and begins its spiral course around the North Centre. Here its expanding tendency is checked, and it begins slowly to wind around its spiral path in the North, going round many times from H to K, and arriving finally, about June 22nd, on the inner circle of the Tropic of Cancer, as represented by the circle L M O P.

By uniting Diagrams I. and II. a similar explanation may be given of the sun's crossing the equator from the North to the South. Thus we have followed the sun in his path for one whole year; and we find that the secret of its motions must be represented, not by one spiral alone but by TWO, as has been already indicated in the figure 8.

Many questions will doubtless arise in the minds of those readers who have carefully followed us thus far; and some of these questions we may consider in further articles. But before closing this chapter we would remind our readers that we have kept principally to the facts of the case. It is a fact that the sun circles (or "spirals" if we may coin the term as a verb) round the North for six months of the year. And the evidence we have given shows that it is also a fact that the sun for another six months "spirals" around the Southern Centre. And we Zetetics know that the surface of still water is absolutely level, and the land therefore generally horizontal. We have accordingly attempted to coordinate these grand facts, and we believe that such an attempt is now recorded in print for the first time in the history of Zeteticism.

# CHAPTER III: ELECTRICITY AND MAGNETISM
## *versus* The Hypothetical "Attraction of Gravitation"

Scientists have long desired to find a physical basis for that which they are pleased to term "the Law of Universal Gravitation." Much better would it have been if they had first sought proof as to whether universal attraction is a fact, or only a mere theory. In many cases the phenomena on which they rest their theory are capable of explanations apart from that theory.

That bodies in some instances are seen to approach each other is a fact; but that their mutual approach is due to an "attraction," or pulling process, on the part of these bodies, is, after all, a mere theory. Hypotheses may be sometimes admissible, but when they are invented to support other hypotheses, they are not only to be doubted but discredited and discarded. The hypothesis of a universal force called Gravitation is based upon, and was indeed invented with a view to support another hypothesis, namely, that the earth and sea together make up a vast globe, whirling away through space, and therefore needing some force or forces to guide it in its mad career, and so control it as to make it conform to what is called its annual orbit round the sun! Theory first of all makes the earth to be a globe; then not a perfect globe, but an oblate spheroid, flattened at the "poles"; then more oblate, until it was in danger of becoming so flattened that it would be like a cheese; and, passing over minor variations of form, we are finally told that the earth is pear-shaped, and that the "elipsoid has been replaced by an apoid"! What shape it may assume next we cannot tell; it will depend upon the whim or fancy of some astute and speculating "scientist."

All this of course is said to be due to Gravitation! We have long since given up the theory of gravitation; in fact that theory went with the globular theory which it was invented to support. We think that the phenomena of celestial motion can be explained by Electricity and Magnetism without having to resort to the theory of universal "attraction" of bodies for each other; especially attraction at such enormous distances as the astronomers postulate. In short, Zetetics agree with Sir Isaac Newton, that "action at a distance" is impossible without some connecting medium: and that, therefore, bodies at a distance can only act upon each other through the ether, and the electric and magnetic currents which are set up in that subtle substance.

The action of the magnet is, however, supposed to be a proof of the possibility of two bodies "attracting" or pulling each other together from a distance; but when this proof is examined it will not bear this interpretation. If we stand on London Bridge we may sometimes see a boat approach the bridge, by the mere action of the wind or tide. It would be highly unphilosophic to say that the bridge "attracted" the boat; and it is equally unphilosophic to say that the magnet "attracts" the needle or any other body. As the boat is carried towards the bridge by the action of the tide, or the currents acting directly upon it, so the needle is deflected towards the magnet by the magnetic currents which act upon it. The magnet, because of its internal arrangement, simply has the power to decide the direction of those currents.

When Mr. Adams, or Le Verrier in 1846, discovered the unknown planet Neptune, through the perturbations of the neighbouring planet Uranus, it was, therefore, no proof, as is commonly supposed, of the universality of the Law of Gravitation; for the perturbations of Uranus might be accounted for by electric currents set up between the two planets as they approached each other. If we were to sit in the telegraph office on this side of the Atlantic, and watch the perturbations of a magnetic needle when a message is being sent across the water, it would not be considered very scientific or philosophical to suppose that some needle on the further side of the ocean was "attracting" or "pulling" at the needle on this side! Would it? It would be a much simpler explanation of the phenomenon to say that the magnetic currents set in motion on the one side affected the needle on the other. This is the explanation respecting currents on the earth; and it is the explanation which is given in the case of "wireless telegraphy." But when the philosophers get among the stars with their supposed immense distances, they have to conjure with the word "Gravitation," in spite of all its infinite perplexities, to account for a simple phenomenon. Is this scientific?

A book has lately been published, entitled: *Aether and Gravitation.* It is a suggestive and well-written book; but before trying to find out either the cause of gravitation, or its basis as a universal law, it would have been better to have examined whether there really exists such a universal force of attraction, or "pulling together" of particles, as is so commonly assumed. If Mr. Hooper's book proves anything, it really proves that there is no need for any such theory of gravitation; and it may be possible that he has intended to prove this, while at the same time using the old terms or phrases connected with that theory, so as not to excite the opposition of scientists who are still wedded to such an unphilosophical notion as "action at a distance." But this theory, at the outset, is taken for granted, as is also the globular theory of the earth and its supposed motions. In fact the author in another place seriously sets himself to enquire as to what is the "cause of the earth's diurnal motion"!

Would it not be more logical to first enquire whether the earth really has any such motion? We think so. Astronomers have long been puzzled to account for the earth's supposed diurnal motion. They have no idea what causes it. A primitive impulse will not suffice, as it would require a continued and continual impulse to equalize the "attraction" theory: and so they have invented what they call a centrifugal as well as a centripetal force. But these "forces" only exist in the brains of astronomers and their disciples. It would puzzle the wisest of them to give an unanswerable proof either that there are any such "forces," or that the earth has any diurnal or orbital motion arising therefrom.

Both of these unproved and unprovable theories hinder Mr. Hooper from coming to right and logical conclusions, and so they spoil his book. These theories have beclouded the brightest intellects which have tried to solve the "riddle of the universe." Zetetics want something simpler, some thing more in harmony with facts, experiments, and general observation; and we are persuaded that the connected and kindred forces of Electricity and Magnetism afford to us all the proof which we need.

## ELECTRO-MAGNETISM

The forces of the universe are one: or rather, they are derived from one source, and so are transmutable. They are therefore practically the same, whether applied to things terrestrial or things celestial. To illustrate these we will quote from a current number of the Tramway and Railway World, in an article under the above heading. The article of course deals with the practical application of Electricity.

### DIAGRAM IV

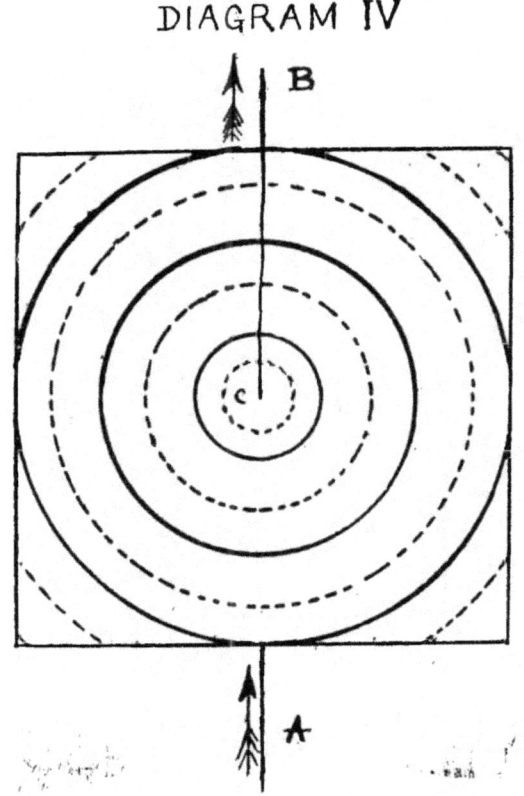

"When a current flows through a straight wire A B, [diagram IV.], a magnetic field is produced around it. The character of this field is shown in the figure over the case when the current is flowing upwards through a vertical wire. When the current is flowing downwards, the field is of exactly the same character, except that the lines of force run in the opposite direction round the wire."

DIAGRAM  V.

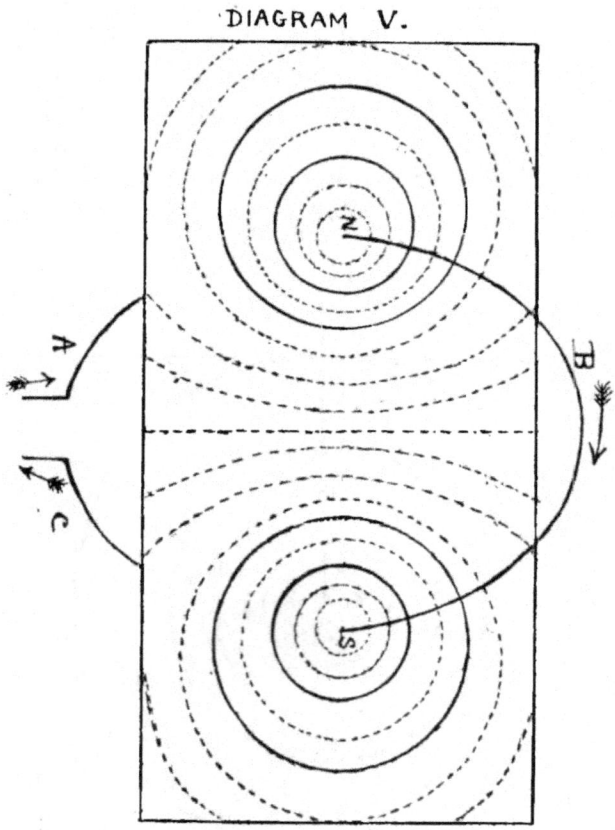

Now by reversing the current as is stated above, we get a similar magnetic field with the lines of force going around in the opposite direction.

We may take diagram IV. to represent the magnetic field in the northern circle, with (c) as the North Centre. But instead of reversing the current and so altering the direction of the circular lines of force, we may take a second field to represent the southern circle with the lines of force going round ABC and so from B to C in the opposite direction, as represented in diagram V. This will illustrate the currents flowing around the South Pole.

In the above diagram (No. V.) a circular current is represented as going along the wire from A, through N, to B in one direction, and so on from B to C through S in another direction. The electric current thus going in an opposite direction through the wire, at S, from that at point N; the circular lines of forces, or magnetic currents, travel around S in an opposite direction to those which travel around point N; and thus we have an illustration of the two great currents which circle respectively around the North and South magnetic "poles."

These great currents meet in a middle and neutral line, or zone, called the Equator, and interlock like the cogs of two connected wheels working together in harmony. This we will illustrate in Diagram VI., in our next chapter.

# CHAPTER IV: TWO "POLES" AND TWO VORTICES

We have seen, from the foregoing evidence that there must be TWO "poles"; but while we as Zetetics admit this fact, we still deny that these "poles" are such as would be required by a globe at each end of its supposed polar axis. They are simply magnetic poles like the "poles" of an ordinary magnet, and not the poles of a rotatory sphere of any kind.

It may be objected that the earth's magnetic poles do not quite correspond with the celestial poles. True! but this may arise from the fact that the celestial and terrestrial fields of magnetic operation, though generally the same, are not quite coincident. But this difference may be easily accounted for. There are two celestial poles and two terrestrial poles, and the fact that these poles are alike magnetic, will account for their slightly different positions; as also for the fact that these poles gradually alter their geographical areas. But we cannot at present enter further into this interesting question. Nature seems to work in pairs. We have heaven and earth; the sun and moon; man and woman; positive and negative; North and South; electricity and magnetism, etc., etc.

It is generally acknowledged by scientists that the earth is a vast magnet. Being such, it will naturally have two magnetic "poles," one in the northern circle or circuit, and another in the southern circle or circuit. In the centres of these vast circles of forces are the so-called "poles," one positive in the North, the other negative in the South. The differences between the positive and the negative may in some measure account for the differences of climate in the two circles ; the differences in the flora and fauna; and even, to some extent, for the differences found in the animal world generally, and the various races of mankind in particular. It is a noteworthy fact that God created and primarily placed the human race in the northern circle.

The sun is a vast electric body, circling around and over the earth. And its motions may be known by experiment and observation. It is well known by those who study the laws of electricity and magnetism, that if an electric current be made to circle round any body that body becomes magnetic by induction. Thus in the daily revolution of the sun around the earth we have a physical cause which accounts for the magnetism of the earth. If the sun were to leave the heavens, and cease revolving, the magnetic currents of the earth would rapidly die out. Furthermore, if we were to seek for the physical cause of the daily revolution of the sun, we should find that it is bodily carried round by the Ether, which is in rapid and continual circulation over the earth. Thus the Creator has solved the problem of perpetual motion, or rather has given us an example of perpetual motion, the solution of which man has hitherto tried in vain to discover. If ever it be found out, by man, we may venture to predict that it will be found only in connection with some circulating electric body, or a circulating current of electricity inducing magnetic action.

This circulation of the Ether was advocated in The Earth for September, 1900. The ether was there said to be in "a state of constant flux, like a great stream continually going around the North Centre, or so-called "pole," carrying all the heavenly bodies with it at various heights, according to their varying densities." In fact it was represented as a sort of vortex motion.

## VORTEX MOTION AND GRAVITATION

When Sir Isaac Newton suggested, or invented, or formulated the idea of universal gravitation, eminent mathematicians and philosophers opposed the idea, and suggested certain theories of vortex motion, or motions, around a number of vortices to account for celestial phenomena. And this was done too with the idea of displacing the theory of gravitation.

Our space will only allow us barely to mention such names as Kepler, Descartes, and Huyghens. But these men, while on the right track, if we may so speak, were weighted and hampered by the incubus of the globular theory ! Once vve have proved, by practical experiment, that the earth is a plane, and a vast magnet, then we are open to receive a better idea of vortex motion, and the untenable nature of the gravitation theory. But, coupled with the lately proved fact that, the sun, for at least six months of the year, circles around a southern "pole," we are now compelled to admit that there are two ethereal vortices instead of one ethereal vortex as we formerly supposed. But the subject is too great to be encompassed within the limits of one or two articles. We can at present only indicate the general outline of these forces or causes of celestial motion. We recommend Zetetics to study the kindred sciences of electricity and magnetism if they would arrive at a clear conception of the motions of the heavenly bodies.

Speaking generally we may state that the stars in the southern circle move in circles round the southern centre; while those in the North move around the North Centre. The sun, moon, and planets move in spirals which are almost, but not quite, circles. Coining a word we may say that those bodies which are called "planets," or "wanderers," sometimes "spiral " round the South Centre, when they have south declination. They all revolve much in the same way as the sun, the motion of which has already been explained. There are, however, some peculiarities or special characteristics in their orbits, especially in that of the moon, which will take two or three chapters to go into more fully, and so we must leave the moon's motions for future consideration. At present, proceeding with our explanation of the two vortices, we will further illustrate them by showing the action of two cogwheels when working together in harmony.

This diagram, No.VI., showing two cogwheels interlocking or working together on their respective shafts or centres, may roughly illustrate the action of the two great currents or sets of currents, flowing round the north and south centres like two great vortices. The upper wheel may represent the circles or lines of forces revolving around the North Centre, as represented by the belt A B C. This great belt will represent the magnetic force revolving around the earth between the Equator and the Tropic of Cancer; while the lesser belt G H may represent similar forces revolving around the Arctic Cirle. The lower wheel illustrates similar lines or belts of force revolving around the southern centre; the broader belt, D E F, showing the position of those forces between the equator and the Tropic of Capricorn, and the lesser belt J K those about the Antarctic Circle. The spokes of the wheels may represent meridians, which according to the testimony already adduced must converge south of the equator as well as north.

DIAGRAM VI

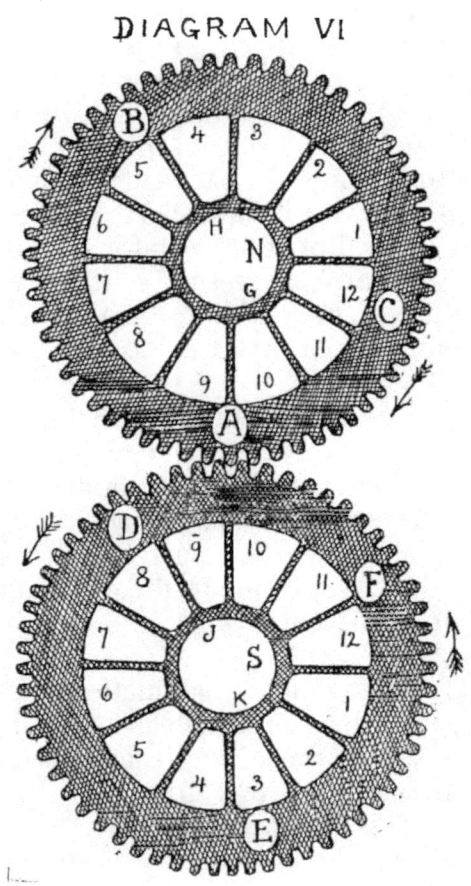

Of course, in the case of the two wheels, the cogs would be rigid, being made of solid wood, or metal. But the great magnetic currents are not rigid, but play into one another at the points of contact about the so-called equator. This has already been explained in a previous chapter on the sun's motion. The sun being a comparatively small, light, electric body, is carried around by one set of currents in the south until, partly by its momentum and partly by magnetism, it is projected into the other set of currents in the North.

Now if the northern currents go round in one direction (the direction indicated by the arrows) it is clear that the southern currents must go round in an opposite direction, so that the two currents may work together in harmony. And this agrees with the facts previously brought forward from the testimony of those in the South. If a spectator in the North looks towards his centre he will see those celestial orbs which have North declination, rise in the East on his right hand. But if a spectator in the Southern Circle looks towards his (S) centre, he sees the southern constellations and the sun rising in the east, but on his left hand. Thus the two vortices work together in harmony, as we have illustrated in the cog wheels.

Respecting the Ether we may briefly state that our own conclusions are, that all motion can be traced to the circulating movement of the Ether; and that the Ether is therefore substantial, or, in other words, a finer form of matter—something approaching the spiritual, if we may apply a material term to such a refined substance. And this agrees with the fact that God created light, or the luminiferous Ether, before He made the sun.

We have generally been taught that matter exists only in three forms ; the solid, the liquid, and the gaseous. But there may be a fourth form of matter, as much finer than gaseous, as gaseous is finer than the liquid; and this form we believe to be the luminiferous Ether. This Ether, being a form of matter, has many, though perhaps not all, the attributes of ordinary matter, such as elasticity, density, motion, and momentum. It thus *carries* around with it all the heavenly bodies at their various distances from the centres.

According to the popular theory of gravitation, these bodies have to pass through the Ether, or some such substance, which is said to fill all "space." But this substance however rare, would continually impede their progress, and would ultimately bring the universe to a standstill! And so the inventors of the gravitation theory had to invent a sort of "frictionless Ether" to allow all their heavenly bodies to pass through it in constant periods of equal times. But a frictionless substance would be no substance at all; and the idea is contrary to all our experience and experiments.

The truth does not require such contradictions and absurdities. The Ether which fills all space from the earth to the firmamental vault above, being enclosed in that vault, circulates, and, being substantive, it carries around with it all the heavenly bodies in their various orbits, which are more or less circular. These bodies have different times of revolution according to their varying heights; which heights are regulated by their varying natures, sizes, and densities. The sun and moon, which are by far the largest of these celestial orbs, are not more than about 30 miles, or about half a degree, in diameter.

The highest of these floating orbs is probably not more than some six or seven thousand miles high. They are all for the most part mere "lights," or more literally *light holders* (Gen. i. 14-16), *i.e.,* centres of electric, or magnetic forces. These forces radiate from them and affect the atmosphere, and whatever comes in contact with that atmosphere, or whatever breathes it. But this opens up too vast a field to be treated of here.

We trust we have given sufficient evidence to show that there is good cause why Zetetics discard, altogether discard, the modern theory of universal "attraction," an unavailing pulling and tugging of all bodies to get together. We think we have shown that there is no need of such an absurd hypothesis in connection with the Plane Truth; but that, known and practical forces such as Electricity and

Magnetism are quite sufficient to account for all celestial phenomena.

# CHAPTER V: "DEGREES"

Leaving the question of the path of the sun as a luminous orb, we may go on briefly to consider the further question of the motions of light, or the path of the sun's rays.

Even our opponents must admit that there may be a difference between the path of a luminous and moving body, and the path of the rays of light which flow from that body. In this connection we shall have to consider the question whether rays of light move in straight lines or in great curves. Common opinion asserts that the former is the case, whereas it has been shown that the latter is the truth. (See *The Earth* for May and June, 1901, under the heading *Direction of Sunrise and Sunset*.) This is mentioned to show that we are starting no new idea to account for southern phenomena. Electricity behaves in a similar manner both north and south of the equator. Recognizing that electricity and light are simply two forms of one force, we shall first proceed to say a few words about "degrees," and then try to answer an objection which may be raised against what we have already advanced respecting the two vortices and the equatorial figure 8.

The term "degree," or "degrees," is used in a variety of senses, but the primary meaning is that of a "step or grade in progression," whether of rank, dignity, or distance; or the divisions of a circle whose circumference is divided into 360 parts. The latter is the meaning intended here. But we have to enquire what it is that so divides the circle, and what circle it is which is so divided; and, in the case of latitude, whether there be a circle or not. We see what are called degrees marked on a globe, and we are led to conclude from astronomical works that degrees are necessarily connected with a spherical body. But there could be degrees on a plane surface, or the divisions of a circle lying in a horizontal plane, and the "degrees" lying along the radius or diameter of such a plane circle, which of course would be no part of a circle. So that, *prima facie*, "degrees" are no proof of the globular theory.

We ought further to enquire what it is that makes these degrees. The astronomer makes lines on the globe, the geographer puts them on the map and conforms his lines of longitude to the globular theory; and, rightly or wrongly, he makes the general outlines of his continents conform to these degrees. But what we have to ask is: what is it that constitutes a degree in Nature—out of doors in the open, not in the study of the astronomer?

In pursuing this enquiry we shall find that a degree is dependent upon, and is measured by the position of rays of light; that is, rays of sunshine. The position of these rays must be dependent upon the nature of the motions of light, as well as upon the actual position and motion of the sun in the heavens. Now we know that the direction of a ray of light depends upon the medium, or the media, through which it passes. Over short distances on the earth it is found that rays of light travel practically in straight lines while passing through a medium of uniform density, say along and just above the surface of a canal where the atmosphere is of the same general density throughout its length. But we know also that the density of the atmosphere varies and lessens as we ascend ; in other words the density of of the atmosphere increases from above downwards. A ray of light, therefore, coming down from above through media of varying density is subject to certain conditions differing from those which prevail for a ray of light which passes through a horizontal medium of uniform density. But more of this anon. We will first consider what are called

## "DEGREES OF LONGITUDE"

Longitude is the distance East or West from a given meridian. A line drawn from the centre, or "pole," to the outer circumference, or equator, is called a meridian, and represents all places which have the noon-day sun on the mid-heaven at the same time. The word comes from a Latin word, *meridies*, which signifies the middle of the day. This may be illustrated in the following diagram (No. VII.),

where the line C D would be one of the twelve meridians there
shown.

DIAGRAM VII

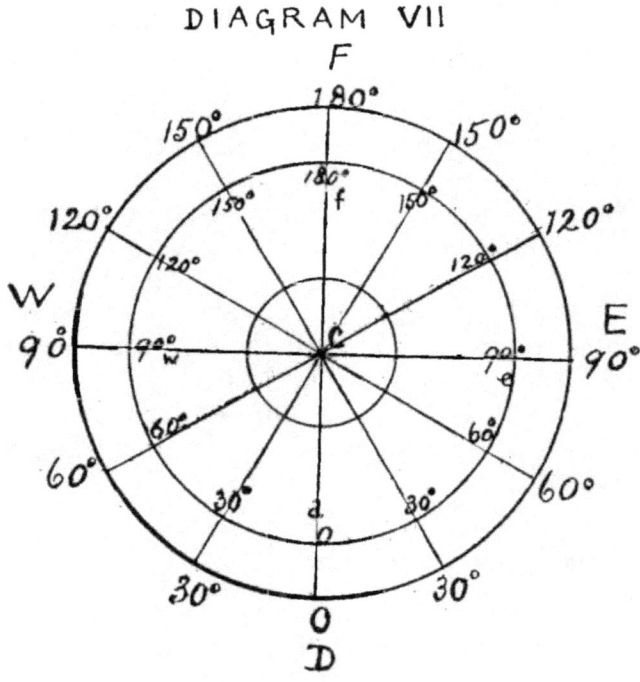

In the above diagram the outer circle is divided into twice 180°, or
360°, dividing the circle into twelve parts: and, reckoning from any
given meridian, say C D running through Greenwich, each line
would represent 30° from the preceding line. Thus, reckoning from
D towards the left hand, we should say 30° West of the prime
meridian C D; 60° W; 90° W, and so forth up to 180° W. And
following round the circle on the other side we should similarly
reckon as so many degrees to the East, until we arrive again at 180°,
when East and West meet from a given meridian.

It would be interesting to go into the question as to which meridian
ought to be counted the prime meridian, but we cannot enter into
this consideration at present. Suffice it say, that we think the prime

33

meridian should be for the North, where man was first created, namely, about 45° east of Greenwich, running generally along or near the banks of the great river Euphrates, which passed through Eden from North to South. This is where the Creator originally placed the Day Line.

Referring to diagram VII., if the outer circle D W F E be taken to represent an equatorial circle, the degrees on that circle would each represent about 69½ statute miles, thus giving a circumference of 25,000 miles; but, upon inspecting the diagram, it is manifest that the degrees on the inner circle d w f e, representing one of the tropics, cannot be so large as those on the outer circle, consequently the same number of degrees on this circle do not represent so great a distance geographically as those on the outer and larger circle. So that using the term "degree" in the sense of geographical distance we see that it is a very elastic term representing distances that must continually vary according to latitude, whether North or South. It is well to remember this.

## DEGREES CONVERGING SOUTH

Now as we have seen that the sun revolves around a southern point in the heavens for one half of the year, so we must conclude that during the same period degrees converge south of the equator, as they do in the North when the sun is North. In fact celestial phenomena south are similar to those in the North, according to the latest evidence obtained; and, as Zetetics, we are willing to give place to facts, while maintaining our right to question mere theories. But as vve have, before observed, these motions have nothing whatever to do with the surface shape of the earth and sea, but rather with the question of the sun's motions, and the motions of the sun's light. Zetetics are therefore at liberty to accept the evidence given for the fact of the midnight sun in southern regions, as witnessed by the gallant crew of the *Discovery*, in the late Antarctic Expedition.

But if we are asked by opponents, how it is that the midnight sun can be seen south at Christmas time? the evident and simple answer to the question would be, because the sun goes there at that time of the year! This motion has already been explained as resulting from the two Vortices of the Ether, or the great electro-magnetic currents, which circle around the two celestial "poles." And any amount of converging degrees south will never alter the shape of the Bedford Level Canal.

## DEGREES OF LATITUDE

Before leaving the subject of degrees, it may be instructive briefly to consider the further question of degrees of latitude.

Terrestrial "latitude" is the distance measured North or South of the equator, along one of the meridians, and corresponds with celestial "declination." It seems a pity that two different terms are used for the same general idea, for it is somewhat confusing to young students of astronomy. But distance from the celestial equator is called declination, while distance from the terrestrial equator is called latitude, whereas celestial latitude means something quite different from either, as we may show when we come to consider the moon's peculiar motions.

Referring again to the above diagram (VII.), we must point out that there is a great fallacy underlying the astronomical and geographical idea respecting degrees of latitude. The meridians, or lines, along which these degrees are measured are assumed to be great circles, like that of the great circle D W F E, only these are said to run north and south instead of east and west. But as we have again and again proved that the surface of the sea is level, and the land generally horizontal, the line D d C must represent a straight line, that is, the radius of the plane and horizontal circle D W F E, the whole diameter of which is D C F.

Now the relationship of a circle to the circumference, known by the name of the Greek letter for our "P" which is called Pi(II), is given in mathematical books as 3,1416. This means that the circumference of any circle is a little over three times the length of its diameter. In other words, counting the diameter as I, the circumference is related to it as I is to 3,1416; that is, the circumference is rather more than three times the diameter. But not to encumber the idea with details, we will take the circumference as 25,000 miles about the equatorial circle, then the diameter would be about 7,957 statute miles, and the radius of course about 3,978 miles. But a quarter of a great circle would be 6,250 miles; there is, therefore a great difference between the length of a meridian from the "pole" to the equator on the plane earth, measured as a straight line, from what a similar meridian would be arching over one quarter of a sphere. But as these meridians are said practically to be equal to the meridian on the quadrant—that is, a quarter of the equatorial circle—the conclusion has been hastily assumed, that the earth must be somewhat spherical, or perhaps pear-shaped. But if we have already proved the earth to be a plane, no amount of assumption concerning degrees can make it to be spherical, therefore some other explanation must be sought for the fact, if it be a fact, that degrees of latitude, speaking generally, are nearly of the same length as equatorial degrees of longitude. And we think that the following explanation may be found in harmony with the plane truth, and what has already been evidenced.

The rays of light travelling north or south are subject to conditions different from those which follow the usual course of the sun. In the latter case the light travels in the same direction as the great Vortices already spoken of, while in the former case, not to mention the question of refraction, the light has to travel at right angles to these currents. In the one case the sun itself moves bodily along one of the equatorial circles, as from o to 30°; whereas degrees measured along the meridional line D d C, must have a stationary sun as it were, say at D, while the so-called "degrees" are measured along the straight line D d C. This may be illustrated by the following diagram.

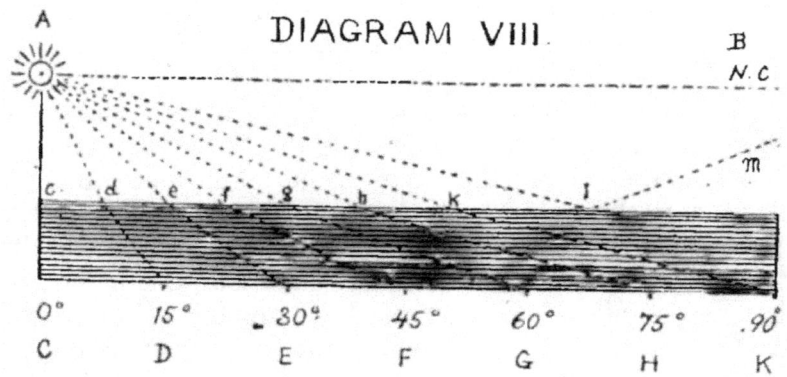

In diagram VIII., A represents the position of the sun when on the meridian for all places along the straight line D d C in diagram VII. This line is represented in the above diagram (VIII.) by the straight line C D E F G H K, with points in the meridian 15° apart. The rays of light from the sun which fall perpendicularly from A to C will of course be without refraction, but as the rays travel further north they fall on the atmosphere at continually lessening angles, and so are subject to greater and greater refraction, until they fall at so small an angle, as at l, that they are deflected off the atmosphere towards m. This is like the action of a stone when thrown almost parallel with the surface of a lake; it skims off the water again and again until its force is spent. Such rays as do not enter the atmosphere give no definite light, and so the sun's light, as it recedes north, is gradually dispersed and finally lost at the North Centre. Similar conditions apply to the South.

So the sun is not seen more than 90° either north or south of the equator at the times of the equinoxes, whereas it is often seen at moie than 90° away from the observer on the equator, when it is travelling West around one of the circles with North declination.

Besides, the above diagram shows how degrees, travelling along a straight line, or meridian towards the North, may be deflected so as to lengthen in the North, having more atmosphere through which to pierce, and so being subject to a greater degree of refraction. And this, we are informed, is in harmony with experiment and facts connected with the measures of such degrees.

Inasmuch as astronomers, and surveyors generally, acknowledge that there is a little flattening towards the "pole," so must we be thankful that they admit the earth is flat somewhere! But on the globular theory there cannot be the same lengthening of degrees in the extreme North, as anyone may see for himself if he will draw a diagram of the globe, with a relatively small sun on the equator at the distance of about one half of the diameter. We have before proved the sun to be a small body, not more than half a degree across, and therefore comparatively near the earth.

# CHAPTER VI: DIRECTION OF SUNRISE AND SUNSET

We have now to consider the question of the motions of the sun's light, as distinct from the actual motions of the solar orb.

As we have already intimated, there is a difference between the motion of light from a moving luminous body and the rays of light which flow outwards from that moving body.

It has generally been assumed in astronomical works that the rays of light coming to us from the sun, speaking generally, move in straight lines from that body to us; though they allow for some little refraction when those rays enter the atmosphere. Often we find in their illustrations, that the sun's rays are drawn from the sun in parallel straight lines right to the earth. Because light over short distances moves in apparently straight lines on the surface of the earth it is assumed that the rays of the sun must move in straight lines down to us from that body, situated as it is above the atmosphere. But we can easily prove that such assumption is fallacious.

Place a long rod, or stick, entirely under water, and the stick will appear quite straight. But place the same rod half under water, and half above it and the rod will seem bent or broken at the point of contact with the water. This shows that when a body passes, or rather when the rays of light from a body, pass through a medium of uniform density, no bending or curvature is visible; but that when similar rays of light come to us through media of varying densities, or through the same medium with changing density, the bending or curvature of the rays must take place.

Now the upper parts of the atmosphere, as we have already observed, are much less dense than the lower parts, therefore a ray of light coming down to us from above is refracted more or less out of a straight path. And light, like all other subtle forces, always takes the line of least resistance. Thus, referring back to Diagram VIII. in our previous article, the spectator at the "pole" K, if such a person could be placed there, would receive the sun's rays so much bent, as we have there shown, that the last rays of light from the sun coming to the spectator's eye would seem to come in an almost horizontal direction; and as we always "locate" an object in the direction of the last rays, the observer there would see the sun low down on the horizon, even though at the same time it were in the mid heaven to a person on the equator at C. It is manifest therefore that an observer would not see the same focussed image of the sun at K, as would the observer at C. This fact should prove to us that the place of the focussed image of the sun depends partly upon perspective, due to the position of the spectator, and partly on the condition of density and the amount of the atmosphere through which the rays have to pass. For instance,the rays of light have much more atmosphere to pass through to a spectator at K, than they have in passing to a spectator at C. So the sun appears to be setting to the one while it is high in the mid-heaven to another.

Now as the sun's image appears to set it is often magnified by the particles of moisture which are always more or less present in the atmosphere. This fact is well known to those who study the science of optics; yet objectors to the plane truth often urge the silly objection that the sun's disc ought to appear less when it is further away from us! This, no doubt would be the case if the sun were a non-luminous body. But as we have just observed when the sun passes away from us to the west, its rays of light have to pass through a much thicker stratum of atmosphere, or even through different strata overlying each other, all containing thousands of minute globules of water, which tend somewhat to magnify the sun's disc, and so prevent the perspective foreshortening of its diameter which would otherwise ensue. In fact when there is more than the usual amount of water in the atmosphere this diameter seems largely increased rather than diminished when the sun is setting. So much for that objection.

But we shall have another and more forcible objection to meet, now that we have advanced our new explanation respecting the equator.

If the sun crosses the equator in the manner described in the previous chapters, it may be asked: how is it that the apparent position at sunset or sunrise is only very slightly altered on the day when the sun alters it declination from North to South? We will try to meet this objection, and by way of illustration we shall have recourse to the following diagram (IX.)

DIAGRAM IX

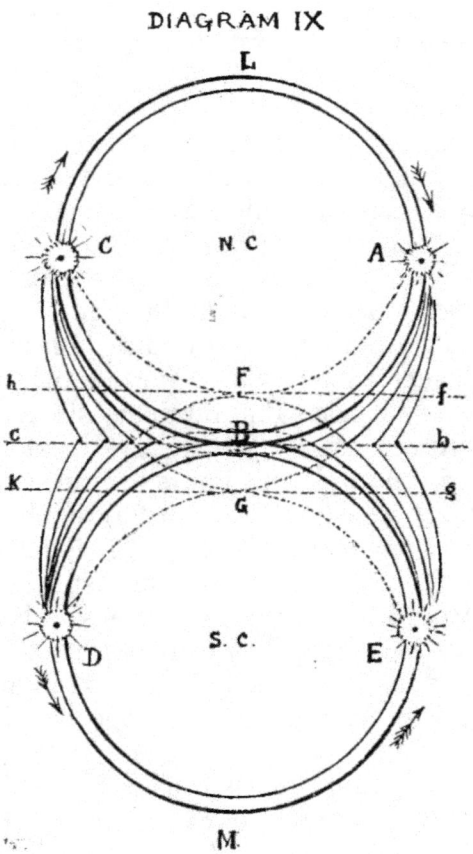

In the above diagram , N C may represent the position of the North Centre, about which the sun revolves when it has north declination, that is from March 21st to September 23rd in each year: and S C the

South Centre, about which the sun revolves during the other six months of the year.

We have already shown how the luminiferous Ether circles about both centres, producing the vortices of which we have already written. These vortices carry round with them the rays of light, as well as the actual body of the sun. In fact a ray of light, or a pencil of rays of light, is simply due to particular undulations of the luminiferous Ether. But let us first consider the action of the vortex in the North Circle.

We will take the time when the sun is said to be on, or above, the equator, A B C L. At these times of the year, March 21st and September 23rd, the days and nights are equal: in other words we have twelve hours daylight and twelve hours darkness. Now we know the sun goes round the circle in twenty-four hours. We also know that, to a spectator, say at B, on the equator, the sun appears to rise due east of him—that is in the direction of B b. But at the time of sunrise to such a spectator, we also know that the sun itself is really at A, because it takes that luminary six hours to go from A (the place of sunrise) to B, directly overhead at noon to the spectator at B; and it occupies six hours for the sun to go from B to C, the place of sunset. These are facts, just as much as it is a fact that water is level and the earth a plane. But it is also a fact that the spectator at B does not look for the sunrise in a north easterly direction, as he would have to do if the rays of light came to him directly in straight lines from A to B. If his face be turned towards the North Centre he sees the rays of light come to him, as he thinks, in a straight line due east on his right hand as from b.

How is this? It is because the rays of light do not travel in straight lines but in great and varying curves around the northern vortex. These rays follow the general direction of the equatorial currents from A to B. The spectator sees them , when they last meet his eye, in the direction of (b) where the image of the sun is necessarily formed for him, and so he thinks the sun itself is actually rising at (b), whereas it is only the sun's focussed image, and the sun itself is at A.

This will further appear if we take the position of another spectator, say at F 23½ deg. North of the tropic of Cancer. The spectator at F, looking, towards the North, does not see the sun rise at his north-east, as it would if the rays of light came to him from the sun in straight lines. Like the other spectator at B, he sees, at this particular time of the year, the sun rise "due east," on his right hand, towards (f); and so to him the sun's image is focussed at (f). This is a fact admitted by all who have studied this subject. And for similar reasons a spectator at G in southern latitudes, also sees the sun rise directly to his east in the direction of (g) on his *left* hand, if he be looking towards his southern centre.

But the actual body of the sun cannot at one and the same time be on (f), and at (b), and at (g); for it is contrary to all the known laws of philosophy for a body to occupy different and distant positions at one and the same time. It cannot, therefore, be the actual body of the sun which these differently placed spectators see in diverse positions, but the various focussed images, which take up their respective positions, according to the various situations of the spectators, and in harmony with the revolving motions of the vortices, and the refractions to which the light is subject in passing from a luminous body *above* the atmosphere through a medium of ever varying density.

But to proceed further to answer the supposed objector's question, why, when the sun crosses the equator from the northern circle into the southern circle, is there so little alteration in the apparent position of sunrise and sunset?

Let us take the time when the sun is said to be on, or very near, the equator in the North. We ay notice that the equator is not "an imaginary line," if we only consider the matter. It is a BELT at least as wide as the sun's diameter; that is about half a "degree," or 32 geographical miles. Within this belt the sun's rays will fall perpendicularly over an area equal to its diameter.

Now take the day before the sun "crosses the equator" from North to South, say the 22nd of September. The sun rises that day, to a spectator at B, in the direction of (b) as we have just proved. It then sets at (c) to such a spectator, that is when the sun has actually reached its position at C, about six p.m. on the evening before it "crosses." The sun goes round to A ith about 24 min. less of North declination each day, rising in the direction of (b); then proceeding to B, overhead, it "crosses the equator" and passes on to D in the southern circle.

"Crossing the equator" is really the sun changing its declination from North to South. Its parting rays come back to the spectator at B, along the curved lines from D. He sees again the sun's focussed image setting at (c) close to the place where he saw it set on the previous evening.

The sun then passes round the southern circle, in the direction of the arrows, until it arrives at E, twelve hours after it was lost to the spectator at B. This observer now sees the sun rise in the direction the light comes to him along the curved lines from E to B, and as the object is always pictured in the direction of the rays last entering the eye, he again sees the sun (that is the sun's focussed image) rise in the direction of (b) near to where he saw it on the previous morning, with only the alteration of the sun's declination, that is about twenty-four minutes of a degree for a day.

Afterwards, the sun increasing its southern declination, that is revolving in its fine spiral nearer and nearer to the southern centre, the spectator in the North sees its rays rise and disappear farther and farther South of him to the S E and S W, until after six months it comes back on the southern equatorial circle, and passes again into the northern circle, thus making in its actual motions the mysterious figure 8, by the two adjoining circles.

Thus we may have a clearer understanding of the Psalmist's expression, which we referred to in a previous article: "His (the sun's) going forth (not the earth turning its axis that way) is from the end of the heaven, and his circuit unto the ends of it." So that there are not only "ends" to the earth (the land) but "ends" to heaven.

The more light we obtain respecting the facts of Nature the better we shall see that Bible statements are harmonious therewith. We cannot, therefore, do better than close this present article with a further quotation from the inspired Word.

"The works of the LORD are great, sought out of all them that have pleasure therein.
His Work is honourable and glorious, and His righteousness endureth forever.
He hath made His wonderful Works to be remembered.
The LORD is gracious and full of compassion."—*Ps.* Xci.

The Sabbath is Creation's memorial.

Again :
"Oh give thanks unto the LORD, for He is good; for His mercy endureth for ever.......
To Him that by wisdom made the heavens; for His Mercy endureth for ever.
To Him that *stretched out* the earth *above* the waters; for His mercy endureth for ever;
To Him that made great *lights*; for His mercy endureth forever.
The sun for the ruling the day (margin); for His mercy endureth forever.
The moon and stars to rule by night; for His mercy endureth forever."—*Ps.* Cxxxvi. 1-9.

Let us say "Amen" to the above inspired words. Even men in high positions in the professed Churches of Christ are now impugning the veracity of the Holy Scriptures, because, forsooth, they cannot reconcile them with modern "science falsely so-called"! Let Zetetics therefore be zealous in upholding the inspiration, and the authority of the Sacred Records. Zeal eventually triumphs,engaging Truths eternally sacred.

# CHAPTER VII: KEPLER'S LAWS OF MOTION

As we have given a general view of the laws of motion affecting the heavenly bodies, it may be well to compare them with those given by popular astronomers.

Let us for instance compare them with the laws of motion given by the famous astronomer, Kepler. By so doing we shall be better able to form an opinion of their respective merits.

Electricity and Magnetism are the forces chiefly required in the Zetetic System to account for the motions of celestial orbs; while the "attraction of Gravitation," whatever this may mean, is needed to give some plausibility for the celestial motions as taught by modern theoretical astronomy.

Sir Robert Ball, is now the chief exponent of the latter system, and he upholds Kepler's ideas respecting planetary motion. Though worldly titles do not necessarily give a man wisdom in the things of God and of Creation, we are willing to speak of him personally with all due respect; but he cannot complain if we attempt to lay bare the inconsistencies of the patchwork system of astronomy which he represents, and which he is paid officially to uphold.

Since *The Story of the Heavens,* by Sir Robert Ball, is one of his most popular works, we shall make some reference to it, and some extracts from it. Sir Robert Ball admits that the ancient philosophers thought that the earth was without motion, and that it was the sun and moon with the stars which revolved around the earth, and not the earth around them. Coming down to the time of Ptolemy, he says of this great astronomer and astrologer:

"The earth according to him was a fixed body, it possessed neither rotation round an axis nor translation through space, but remained constantly at rest in what he supposed to be the centre of the universe. According to Ptolemy's theory the sun and moon move in circular orbits around the earth in the centre."

"Although the Ptolemaic System is now known (thought?) to be framed on quite an extravagant estimate of the importance of the *earth* in the scheme of the *heavens* (!), yet it must be admitted that the apparent motions of the celestial bodies can thus be accounted for with considerable accuracy." (Italics ours.)

We think it is rather to Ptolemy's credit that he did not include the EARTH in "a scheme of the heavens." He did not confound the earth with the *heavenly* bodies as Sir Robert Ball does in his *Story of the Heavens*. However, those who wish to see more of Ptolemy's system can refer to his *Almagest,* written in the second century, which work was considered a final authority in astronomical matters for fourteen hundred years, until the time of Copernicus. We may remark that his great astrological work, *The Tetrabiblos,* is even to the present time reckoned as an authority by those who understand the subject upon which it treats.

But when Copernicus arose, he tried to show that the sun was stationary, and that it was the earth which revolved about the sun, with the "other" heavenly bodies. Of this great astronomer, Copernicus, who gave his name to the modern system of theoretical astronomy, Sir Robert Ball says:

"Copernicus pointed out the fundamental difference between real motions and apparent motions; he proved that the appearances presented in the daily rising and setting of the sun and the stars *could be accounted for* on the *supposition* that the earth rotated just as satisfactorily, as by the more cumbrous supposition of Ptolemy."

"Copernicus transferred the centre about which all the planets revolve, from the earth to the sun."

The latter was no doubt something for any mortal man to accomplish. But it does not seem such a great achievement if he only showed that the movements of the heavenly bodies could be accounted for "just as satisfactorily as Ptolemy accounted for them." And if Ptolemy's explanation was unsatisfactory, what had been gained by shifting or "transferring the centre of motion" from the earth to the sun, we leave for Sir Robert to explain, with some more serious discrepancies which we are shortly about to point out.

## KEPLER'S PROBLEM

After achieving the mighty task of shifting the centre of the universe to the sun, astronomers were still dissatisfied with the circular theory of planetary revolutions. But if the planets (among which the earth was included) did not move in circles, what then was the figure of their orbits? Such, we are told, was the great problem which Kepler proposed to solve; on which the writer of *The Story of the Heavens* remarks: "To his immortal glory he succeeded in solving and in proving to demonstration." Yea, he further says: "The discovery of the true shape of the planetary orbits stands out as one of the most conspicuous events in the history of astronomy." Let the reader remember these words later.

What then is this great discovery? Or, in other words, What is the figure, other than a circle, which is supposed to represent the planetary orbits. The curve is taken from a group of curves found by mathematicians in those obtained from conic sections. If both sides of a cone be cut obliquely by a plane passing through it, not parallel to its circular base, the outer edge of the section will be that of an ellipse. The orbits of the planets which are "supposed" to revolve around the sun are said to be elliptical.

There is an easy way of making an ellipse known to students of geometry. Fix a piece of clean paper upon a board, and fasten two common pins in it, as in Diagram X., say at A B. Then take a loose loop of thread or twine, and stretch the thread with the point of a pencil. Work the pencil round, keeping the thread at a sufficient tension, and the ellipse will be formed. See the following diagram , which is similar to the one given by Sir Robert Ball.

DIAGRAM X

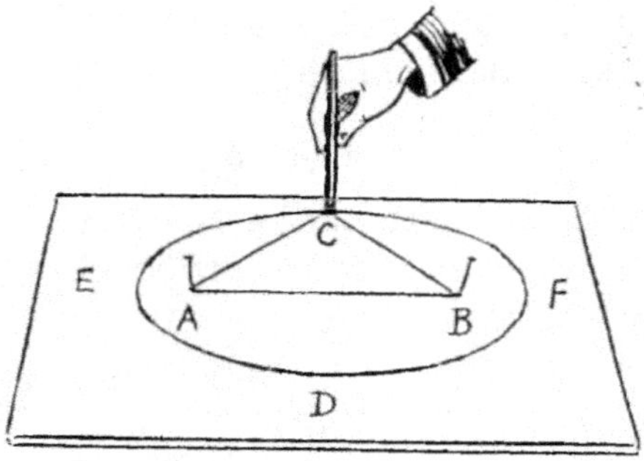

In the above diagram it will be seen that an ellipse is a figure something like a circle pulled out one way, or with its sides squeezed somewhat closer together the other way. It has a longer diameter through E and F, and a shorter diameter through C and D. The places of the two pins A and B are called the *foci*. Every ellipse must have two *foci*. Now we are told that the planets move round the sun in elliptical orbits, with the sun in "one of the foci," say at B. The other focus at A is "to let." That is, there is nothing in it! It has been "to let," like an empty untenanted house, for a very long time. Astronomers should find something to put in it, if only for decency's sake. No mathematician in this world ever made an ellipse with one pin, or one focus; he must use two. But astronomers are a privileged class, and they have to admit that the sun occupies only one of the foci. It never even takes a turn in the other, according to their teaching.

We Zetetics do give the sun a change, for six months, from the northern circuit to the southern circuit. But they keep the sun blazing away in "one of the foci"! If the moon only occupied the other it would not look so empty; but the moon is said to be a lesser planet, and she is wanted with the earth, for the elliptical orbit, and you cannot have the moon in "one of the foci" and in the orbit as well. So the other focus, whichever it may be (they don't tell us which), is "to let." The astronomers do the best they can, and as they have not yet found an occupant for the second focus they say little about it! They might put up a notice board, "to let," instead of which they leave us to believe that something or some body is there.

Now Kepler, being a good mathematician, was of course familiar with the ellipse; as Sir Robert Ball says, "it was to his hand." This was very convenient. Moreover "its properties were known;" this was better still. Then Kepler, as well as Sir Robert Ball would know that *two* foci were needed, not to be left empty and desolate, but to be used by somebody or force in making the orbit. Well, what did Kepler find out? According to the writer of the "Story" (a very interesting "story" too):

"Kepler found that the movement of the planets could be explained by *supposing* that the path in which each one moved was an ellipse. This in itself was a discovery of the most commanding importance."—
*Story of the Heavens*, p. 110

The motions of the planets could be "explained," and explained by "supposing." Then we are gravely informed that this supposition actually "reduced to order the great *globes* which circled round the sun. If the bare supposition above mentioned was of the "most commanding importance," how much more commanding would it be if the mere hypothesis actually "reduced to order" the whole of the solar system? To describe this adequately we should need another degree of comparison above the superlative, something like the fourth form of matter, and we are tempted to ask, Can Nature be pulled about and altered by such hypothetical performances? If so, it beats Joshua's commanding the sun to stand still.

But we shall proceed to show how Sir Robert himself spoils it all, by another supposition given in the latter part of the same work. At present we wish the reader to remember that he endorses, with all modern astronomers, Kepler's Laws of Planetary Motion. He enunciates the first law of planetary motion, which is the basis of the others, in the following words: "Each planet revolves around the sun in an an elliptical path, having the sun at one of the foci."

After giving the above important "law" of Kepler, Sir Robert says, "we are now able to form a clear picture of the orbits of the planets." We hope to compare this clear picture of elliptical orbits with another picture in a future chapter, and a picture drawn from the data supplied us by the great astronomer himself, so that our readers may judge for themselves how much faith to place in the sincerity of modern astronomers, who must know that the ellipse, to them, does not represent what they believe the orbits of the planets, and of the earth, really to be. Kepler was at least honest in his belief.

Before concluding this chapter we wish to point out that the theory of elliptical orbits is closely connected with the theory of "Gravitation," so that if the elliptical orbits be crushed out of shape by further modern theories, the underlying and overlying (also ever lying) theory of Gravitation must go with them.

Sir Robert writes: "Newton's discovery of Gravitation fortifies Kepler's Laws." We shall see. We have, however, again and again pointed out, that Newton never did "discover" gravitation. He invented the theory: or, perhaps it would be more correct to say, he formulated the theory of Gravitation to support the globular theory. But he had not, himself, much confidence in that theory, because it required "action at a distance." This may be seen from his own words. In a letter to Dr. Bentley, dated Feb. 25th, 1692-3, about ten years after his supposed discovery, Sir Isaac Newton makes the following confession:

"That gravity should be innate, inherent, or essential in matter, so that one body may act upon another *at a distance*, through a vacuum, without the mediation of anything else, by and through which their action and force may be conveyed from one to the other is to me *so great an absurdity* that I believe no man, who has in philosophical matters a competent faculty of thinking, can ever fall into it."

And on another occasion, Newton further confesses: "What I call *attraction* may be performed by *impulse*, or by some other means unknown to man." (Italics ours).

Now if we have a "competent faculty of thinking," this passage clearly shows that even Newton's penetrating intellect was unable to frame a satisfactory theory of gravitative action. We have, in a former chapter, given our reasons for believing that the theory of gravitation can be entirely disposed of as an absurdity; and that the force which is needed to account for celestial motion is, as Newton himself suggested, "*impulse*,"—the impulse of the ethereal currents about which we have already written.

Thus we Zetetics have two reasons supplied us by Sir Isaac himself, for rejecting the popular theory of universal gravitation suggested by him . First: T he "absurdity" of believing "action at a distance," with no intervening pulling tackle. Second: That the motions of the heavenly bodies, which he supposed to be due to "attraction," may after all be brought about by the "impulse" of some unknown body or substance. That substance has been found; it is sufficient to carry out the purposes of the Creator, who created it on the first day of Creation week. In the Bible it is called "Light," and in works on astronomy it is called "the luminiferous Ether."

# CHAPTER VIII: ELLIPTICAL *Versus* CYCLOIDAL CURVES

Sanction has been given by all modern astronomers to Kepler's Elliptical Curves as representing the orbits of the heavenly bodies. Many of their books give figures of the Ellipse as representing the orbit of the earth as a supposed planet. Instances abound everywhere, but we have given one from the *Story of the Heavens*. This is enough for our present purpose.

Having shown that Sir Robert Ball endorses the theory of elliptical orbits, we now proceed to refute that theory from data supplied in his own book.

A description of the Ellipse has already been given (see Diagram X.) We m ay give the following definition of a true cycloid: "A cycloid is a figure described by a point in the circumference of a circle which rolls along an extended straight line till it has completed a revolution."

Of course the above is a definition of a perfect cycloid. It would be made by one revolution of the circle, or wheel. For instance, we might fasten a nail so as to project from the felly, or outer rim, of a carriage wheel; then place the wheel against a smooth surface of an upright wall so that the nail (or a piece of chalk if preferred) will mark the wall as the wheel is moving along. The curve of the figure, thus made, will be a cycloid.

This curve has some rather curious properties. Many imagine on first thought that the nail would describe a circle; some have thought of the ellipse; but it would describe neither of these curves. Others again have thought that the nail would move forward at the same uniform rate as the whole wheel or the hub of the wheel, but it would not do so. It would not even always move at the same speed. It would depend on which part of the wheel it occupied, either above the axis or below it. But we may have more to say about this later. In the meantime each reader may try the experiment for himself. If he does so he will better understand what follows. It is evident that if the axis of the wheel be kept stationary the nail or chalk would describe a circle. And this was what the figure of the earth's orbit was first believed to be, when once the astronomers had set the earth in motion, or fancied they had done so, and made the sun stand still.

Then came the ellipse, a sort of elongated circle, a figure having two centres, or rather two foci. But if a circular hoop be elongated to form a sort of ellipse, and if a rod be fastened in one of the foci, and the ellipse moved bodily forward, a nail or piece of chalk placed anywhere in the revolving circumference would never describe an ellipse, or any figure like an ellipse. If this be doubted let the experiment be tried.

Now, as we have said, as long as the sun was supposed to remain stationary in the centre of the earth's supposed circular orbit; or afterwards in one of the two foci of the ellipse; the circular or elliptical orbit might pass unquestioned; but since the astronomers have come to believe, and to openly teach, that the sun itself is moving bodily forward through space, neither the circular orbit nor the elliptical orbits of the planets can be tolerated for one moment. Yet they still profess to support them!

When Copernicus "transferred" the centre of the universe to the sun, and sent the earth (in theory) revolving about it, subsequent astronomers should have left it there, or else have owned that their predecessors had made a great blunder. But instead of doing this they go on adding theory to theory, like putting new wine into old bottles, until the fermentation is such that the old skins will hold no longer and so they burst. This is what has happened to Kepler's elliptical orbits; and of course to those laws of motion which were immediately connected therewith. We now proceed to give the proof of these statements.

The theory of elliptical planetary orbits was, to some extent at least, consistent with the theory of a motionless sun in one of the foci, if only they could have found a tenant for the empty focus. But later astronomers, after Sir William Herschel, have sent the sun off on a tremendous journey through space, towards the constellation of Hercules, the principal star of which is said to be millions and millions of miles away! And, what is worse, the sun has carried away "one of the foci" with it! and is still carrying it away at the rate of about 20,000 miles per hour! Will this not burst the skin of the ellipse? Let us see.

In the *Story of the Heavens* we are informed that Sir William Herschell was the first to solve "the noble problem" as to whether the sun was at rest "in the middle (one of the foci?) of the solar system," or "whether the whole system, sun planets and all, is not moving on bodily through space"? The writer who flatters his predecessor and fellow astronomer by asserting that he solved this "noble problem ," says the latter was discovered to be the fact. He writes:

"Our sun and the splendid retinue by which it is attended are moving in space."

Moreover, he states that Sir William further discovered the direction in which the system is moving; and also the rate of the motion. We had better give the statement in Sir Robert's own words:

"The sun and his system are *now* hastening towards a point of the heavens near the star *Delta* Lyrae. The velocity with which the motion is performed corresponds to the magnitude of the system: Quicker than the swiftest rifle bullet that was ever fired the SUN bearing with it the EARTH and all the *other* planets, is *now* speeding onwards. We on earth participate in that motion. Every half hour we are about 10,000 miles nearer the constellation of Lyrae than we would have been if the solar system was not *animated* by this motion. As we are proceeding at this stupendous rate it might at first be supposed that we ought soon to get there, but the distance to the stars in that neighbourhood *seems* not less than those of the stars elsewhere." Page 429. (Italics ours).

Now we ask what becomes of Sir Robert's endorsement of Kepler's elliptical orbits? Had he forgotten the ellipse theory when he penned the above lines? He is too clever an astronomer to have forgotten such an important matter. Did he then presume on the forgetfulness of his readers, or did he think that they would never see through it?

The ellipse is broken up. The earth cannot revolve in an elliptical orbit "now" that the sun is hastening at a "stupendous rate" towards a distant star or greater sun! The sun in "one of the foci," is carrying it off at a stupendous rate. The ellipse is broken; the skin-bottle has burst; the new astronomical wine has burst the old skin of Kepler's ellipse, and the wine of "gravitation" is all running out. Sir William Herschell has broken to pieces the old Keplerian skin, and another titled astronomer publicly declares the "fact," and rejoices in the discovery!

We find no fault with Sir Robert for his laudatory remarks upon Herschell's noble discovery; but we think he should in all decency throw away the old Keplerian skins "*now*", that he sees they are burst asunder. But he does not. Nay, rather he tries to persuade his readers that the old skin bottles, labelled "Kepler's Elliptical Orbits," are in good condition yet. It will not do. Sir Robert, the old skins are burst! They will not contain your new wine; and if you want to preserve that wine, the wine of the "new astronomy," you must get new bottles, or new skins.

Every Zetetic can see the rent in the old theories, and if others care to examine them we will show them by the following diagram.

DIAGRAM XI

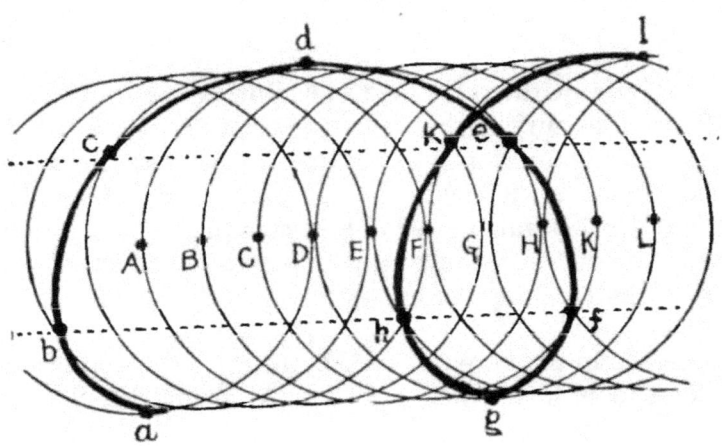

Let A B C, etc., represent the supposed path of the sun in space, as it rushes along towards the constellation of the Lyre, from A towards L. When the sun is at A let the "globe" be at the starting point of its orbit at (a). By the time the sun has gone to B the "globe" will have travelled through about one sixth of its hypothetical orbit, and will have arrived at (b). When the sun has reached the place in its path represented by the point C, the earth will have arrived at point (c). At the end of six months, when the sun has arrived at D, the earth will be found at (d). Two months later when the sun has reached the place represented by E, the earth must have arrived at (e). Similarly when the sun is at F, the "globe," as it is called, must have *preceded* it to and be found at (f). And finally, at the end of the year, when the sun has arrived at G the globe must have completed its revolution, and will be found at (g), ready to commence the cycle again towards h k and l.

Now let a curved line be drawn from the point (a) through the points b c d e and f to g, and we shall at once see the nature of the curve which would be formed by the earth under the above circumstances, that is with the sun travelling along a central line dominating the curve according to the proportions given. The speed of the sun given above, by Sir Robert, is such that nearly twice the distance the earth is said to be from the sun would be covered by the motion "now" claimed for that luminary.

The curve shown by the thick or darker line in the above diagram is a sort of cycloidal curv e; that is a modification of the true cycloid, crushed in at the extremities because the central body has not proceeded as far as it ought to have done to be equal to one revolution of the orbital body. But we leave this fact for the present, and look at the curve produced by taking the proportions given by the two titled astronomers referred to.

Can anyone see in the above curve anything like the ellipse given in the previous chapter, which was copied from the *Story of the Heavens*. An ellipse is formed by one of the conic sections; that is when the cone is cut by a plane passing through both sides, and not parallel to the circular base. Let our readers cut a cone, out of an apple for instance, the astronomical apple of another titled astronomer. A cone is a solid body having a circle for its base, with the top or apex rising to a point above the centre of the base. Let this cone be cut by a thin knife passing evenly through both sides, but not in a plane parallel to the base. If cut in a plane parallel to the base we simply obtain another circle. But cut obliquely across we obtain the ellipse. The ellipse will be the outer edge of the section.

Now whoever in the world obtained a conical section, having for its outline a figure like a kidney, the figure in the above diagram? Would it be possible to cut such an outline in cutting through a cone according to the conditions named? It would not.

It never has been done. It could not be done. It is impossible even for Sir Robert Ball to make a conic section like the above. Yet the above figure represents the orbit of the "globe" according to the latest "science." This is up-to-date "now," though it may not be so in a few years' time. And yet the astronomers keep repeating to us the exploded theory that "all the planets move in elliptical orbits"! Why do they do so, when it is evident that, according to the latest theory, they do not move in closed orbits at all?

Take the curved line from (a) to (g). This represents one complete orbit. From (g) to h, k, and l, is the first half of another orbit. So that there is a great opening in the ellipse equal to about one third of its diameter! Or, in other words, referring to our former figure, the skin bottle which safely contained the old Keplerian wine, has been torn open by the ruthless requirements of later theories; the wine, or system, which it contained, with its sugar of "gravitation," has all run out, and "now" the bottle is lying Hat on the ground, with a great gash in its side. R.I.P.

# CHAPTER IX: THE MOON'S MOTIONS part.1

Logical reasoning is not a prominent feature in scientific works. Either, assumptions are made which are unsupported by facts; or, conclusions are drawn which facts do not warrant. In some instances evident conclusions are overlooked.

Concerning elliptical orbits we have shown that they are inconsistent with modern additions to the astronomical system. Eliminations have not kept pace with these additions, as everyone knows who understands what an ellipse is. Stationary foci are necessary to its completion in a closed circuit; therefore, if we move forward the foci we spoil the ellipse. Even a writer of scientific fiction, as clever as Sir Robert Ball undoubtedly is in this department, must see this, if he would only permit a humble Zetetic to point it out to him. Remember, dear Sir, some Zetetics can reason a little, though they may have neither titles of worldly honour, nor powerful telescopes.

In the last chapter we showed that the supposed orbit of the earth cannot possibly be an ellipse, with the sun in "one of the foci" moving forward at the "stupendous rate" *now* taught by modern astronomers. A rate almost equal to the diameter of the earth's former hypothetical elliptical orbit. In diagram XI. we illustrated what the orbit of the earth would be like with a moving centre, going forward at the rate of speed there given.

We have further to remem er that the moon at the same time is supposed to be going round "the globe" in another elliptical orbit. In fact all the planets and their satellites, as well as the earth and the moon, are all supposed to be revolving round their "centres" in elliptical orbits. No other orbit is supposed to agree with the theory of the "Attraction of Gravitation."

It shows the loose way, too, in which astronomers write and speak of any planet going round a "central" object in an elliptical orbit.

The ellipse is never constructed on its centre, but on the two "foci," which are at some distance from the centre on each side of it. The centre proper would be the point of intersection of the diameters, or the major and minor axes.

Now if the so-called "Attraction of Gravitation" agreed with the supposed elliptical orbits of the planets, what becomes of this hypothetical force when the elliptical orbits no longer exist? Will it now be made to fit in with fresh orbits; those required by the new "science"? It must be a very elastic force, if it can be made to fit in with forces so diametrically opposed as a stationary "central" sun, and a sun running off with "one of the foci" at the "stupendous rate" now adopted! It is remarkable how "scientists" can ignore their own theories, and adopt fresh ones without any signs of compunction or remorse for past "scientific" blunders. Even their former hypothesis was easily questionable. Dr. Owens remarking on it, says: "The late hypothesis, fixing the sun in the centre of the world, is built on fallible phenomena, and advanced by many arbitrary presumptions against evident testimonies of Scripture and reason, as probable as any which are produced in its confirmation." (*Exer.* 36, p. 636.)

But at present we are concerned only with their latest theories, and we are showing the patchwork nature of the new system of astronomy. New patches or new theories are added to the old ones, with which they do not agree, so that instead of mending the "system," or systems (for various systems are in the field) the rent is made worse, and so much worse that its nakedness is "now" uncovered. To bear us out in this statement, we need only refer our readers to diagram XI., in the previous chapter, and compare the orbit of the earth there drawn out to modern additions, with the elliptical theory which was supposed to be in harmony with Kepler's Laws of Motion.

The ellipse is rent asunder, and the gash in its side is as wide in extent as the distance traversed by the sun in its supposed translatory motion through space towards a northern constellation.

Now all this time the moon is said to be revolving around its "centre," the earth, in another elliptical orbit; but if the earth's elliptical orbit has been broken open, and thus spoiled, by the translatory motion of the sun through space, what must happen to the supposed elliptical orbit of the moon, which has to keep up with the earth in its mad career after the sun? It will be doubly spoiled. The ellipse will not only be broken up, but like the child's spring hoop with which it has been playing, or the cyclist's rubber tube which has been smashed through dashing against some (Zetetic?) obstruction, the hoop flies open, the cut rubber bursts apart, and instead of forming anything either like a circle or an ellipse, it wriggles about more like a serpent. The following diagram will illustrate this.

DIAGRAM XII

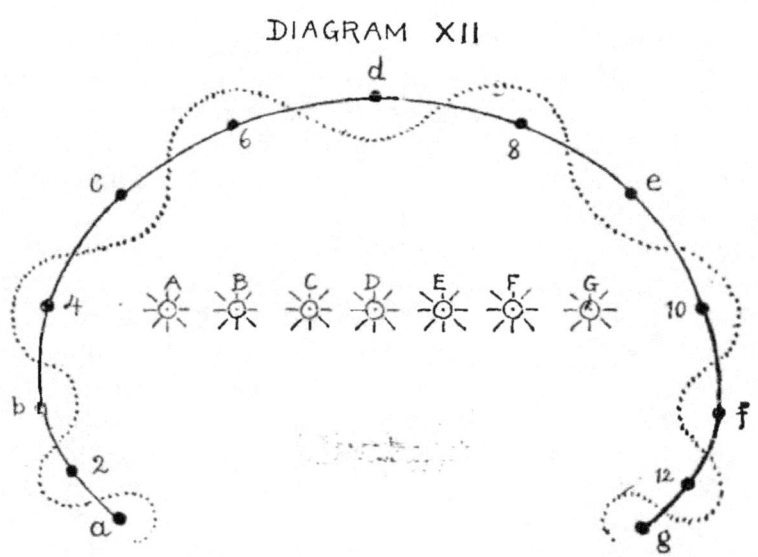

In the above diagram A B C, etc., represents the distance the sun moves in one year. The respective "centres" A B C being two months apart. The small letters, a b c d, etc., represent the respective positions the "globe" must occupy at the same times. That is, when the sun is at A the "globe" will be at (a); when the sun is at B, the sea earth-globe must be at (b); and when the former is at C the latter must be at (c), and so forth.

As the points a b c, etc., represent the earth's positions two months apart from each other, we add the points 2, 4, 6, 8, 10, 12, in between these. The points will then represent the lunar monthly positions of the "globe" in one solar year. Now the moon makes one revolution of the Zodiac in each lunar month. But according to modern astronomical theories, the moon has to keep up to the earth as it revolves after the sun in its cycloidal orbit. To do so it would be compelled to wind in and out, first one side and then on the other of the earth's peculiar orbit, as shown in the figure. What then, we again ask, becomes of elliptical orbits?

How would it be possible, if the moon had such an orbit, for it to pass through all the signs of the Zodiac in a lunar month? It would require two months for the moon to appear to pass through the twelve Zodiacal signs. But we know that the moon passes through these signs in one month, or, to speak more accurately, the twelve signs of the Zodiac each pass the moon in a lunar month, because in their daily revolution round and above the earth they go faster than the moon, as we may show in the next chapter, when we come to treat of the actual motions of the moon as compared with those of other celestial bodies.

Furthermore, according to the motions of the moon as illustrated in the above diagram , that luminary, as seen by an observer inhabiting the whirling globe, if such inhabitant were a possibility, would sometimes appear stationary, at other times "retrograde," and then again "direct." But, like the sun, the moon is never observed to be anything but "direct" in the Zodiac; which is quite sufficient of itself to prove these modern theories of motion to be altogether unreal and fanciful.

A gain, it would be utterly impossible, on modern, scientific theories, to calculate eclipses of the sun or of the moon. The motions predicated of each factor in the problem would be so very intricate. For instance, the sun is said to be moving forward with a translatory motion through "space," towards a particular constellation millions and millions of miles away. Is the sun moving in a straight line towards that constellation, or any particular star? Or, would it not be moving in another "elliptical orbit" round that distant star as its "centre"? If not moving in a straight line it would of course have to carry the "globe," and the other planets, with all their satellites, in a direction different from that which would be traversed if the sun's translatory motion were in a straight line. All these factors, and many others, would require to be properly solved before the relative and actual speeds of the bodies concerned could be calculated; and when these calculations were made on present data, how could the position and time of an eclipse be foretold? It would be impossible.

We might illustrate these by something analogous on earth, and therefore more tangible and intelligible. Suppose an ironclad were sailing round Great Britain, the path of the man of war would be different if this island could assume different shapes. It might be an ellipse or a circle, an oval or a triangle, an irregular figure or a rectangle. Then let us suppose that, circling round the man of war was a small fast steamer plying daily, and keeping at a certain distance, or distances, from the larger vessel. This would represent the "globe" in its supposed journey round the sun.

Then let us further suppose a whale, or some large fish, was constantly swimming "round" the small steamer, keeping at a certain and respectful distance all the time. This might represent the path of the moon, according to modern theories. Now can anyone tell us, what would be the actual path of the whale, or moon? And further, suppose the man-of-war did not always steam forward at the same rate, and the small steamer "circling" round it did not always sail at an uniform rate, and that finally the fish itself was sometimes slow and sometimes fast in its revolutions round the steamer; who could calculate when they would all come into line at certain places, and how long they would be in seeming to pass by each other? No one: especially if the shape of the island around which the man-of-war was sailing, was an unknown quantity.

Yet all these considerations, or similar ones, would have to be taken into account if the calculations of eclipses and transits were worked out according to modern theories. But they are not so calculated, or the Ephemeris would not be so reliable as it is. Yet, shallow objectors to the plane-earth teaching often ask how it is that astronomers can calculate eclipses correctly as to time and place if the earth is not a revolving globe, as they say it is? The answer is, simply that these gentlemen do not calculate eclipses according to their own, and the most modern theories of celestial motion. This is proved from the fact that the earlier representatives of the Copernican system could calculate eclipses, transits, etc., when they believed in a motionless and central sun. It is only quite recently, comparatively speaking, that the theory of a moving sun has been accepted. Yet it makes no difference to the calculations of eclipses, etc.; but it ought to do so, if those eclipses, etc., had been calculated according to modern hypotheses. The fact, therefore, that alterations in these hypotheses have neither prevented nor facilitated the calculation of eclipses proves unmistakably that modern theories have nothing to do with such calculations.

Let us, therefore, hear no more about the accuracy of these calculations as a proof that we are living on a whirling ball, flying through space, while no one can give us the slightest clue as to where we are in space, where we were when we started, and whither we may some day arrive, according to the ever changing theories of our modern scientists.

# CHAPTER X: THE MOON'S MOTIONS part.2: "A ROMANCE OF THE MOON"

*Showing the fallacy of the moon's supposed pathway in the heavens, and her supposed terrific speed! as compared with her real pathway in the heavens, and her real speed.*

Before dealing with the moon's real pathway in the heavens, and her real motion and speed, we must open with an exposition of her *supposed* path or "orbit," and her consequent speed or revolutions according to the fallacious tenets of modern "science."

Professors of the whirling globe hypothesis tell us that "the moon revolves round the earth in nearly a circle (with the earth at the centre) called its *orbit* or path." One professor states: "the moon has several motions, and regarding that by which she revolves round the earth whilst the earth travels round the sun," he says:"this is effected by a cork-screw motion, otherwise the moon would have to travel on one side of her orbit, in a direction opposite to that of the earth. When new or full the moon is on the ecliptic, or has the same level with the sun as the earth. But when seven days old she is above the earth, so that seven days after, when being full, she is below the earth. To understand this, all we have to do is to *imagine* a straight line which represents the path of the earth in her orbit. Then *imagine* that after being new the moon dives below the earth (but still keeping about the same distance from the earth). She then ascends on the other side of the earth and becomes full when she again reaches our level. She is then outside the earth. When new she is inside. This is why the sun can only be eclipsed by the moon when she is new, and also explains that the shadow of the earth can only eclipse the moon when the latter is full."

A complete circle like the annexed diagram , contains 360 angles of I deg.; 4 angles of 90 deg.; and so on.

DIAGRAM XIII

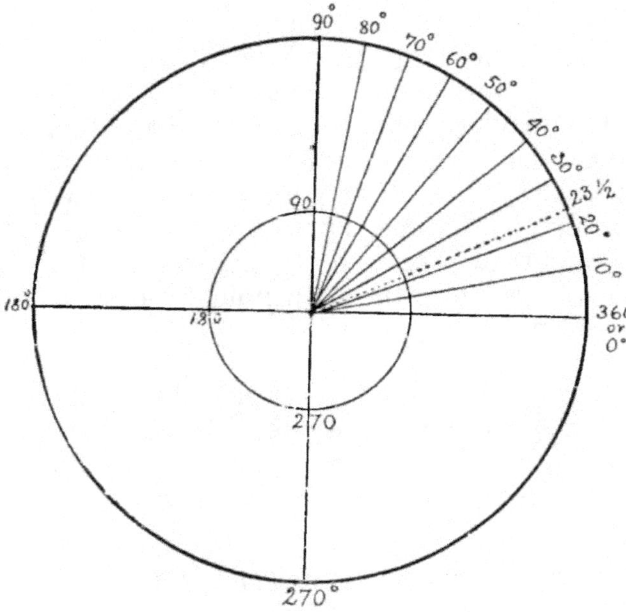

Astronomers conceive such a circle, with its centre at the centre of the earth, and upon this ingenious hypothesis they determine the angles formed by the supposed planes referred to; and they assert that they have thus found that the angle made by the plane of the ecliptic, and the plane of the earth's motion of rotation is 23 deg., or thereabouts (say 23½) and the angle made by the plane of the ecliptic and the plane of the *moon's motion* round the earth is a little over 5 degrees.

The earth's "circumference" is said to be about 25,000 miles, that of the moon 2,160 miles. The earth's rotatory speed, for which they can never find a proof, over 1,000 miles per hour, and its orbital, or rushing round and round the sun speed, about 1,100 miles per minute. The moon's speed round the earth is stated in *Marvels of Science* to be 180 miles per minute. The following paragraph is purely an astronomical hypothesis, and is very suitably entitled *A Romance of the Moon.*

Sir Robert Ball, in his *Time and Tide: A Romance of the Moon,* makes an estimate respecting the moon's moment of momentum, which, says he, "may be regarded as the product of the mass and its velocity, the moment of momentum being found by multiplying the momentum by the radius of the path pursued; yet where the body does not revolve in a circle, but pursues an elliptical path, the moment of momentum is to be found by multiplying the orb's velocity and its mass into the perpendicular from the sun on the direction in which the orb is moving."

In quoting the foregoing, and also the following figures given by different modern astronomers, we deem it impossible to adhere to accuracy. Not only because the distances of the heavenly bodies have been differently rated: but in dealing with

## "A ROMANCE OF THE MOON,"

where veracity is not the ruling principle, accuracy is out of court. But if it were a fact that the moon travels round the earth at the rate of 180 miles per minute, it follows that in a month of 28 days it would have travelled about 7,257,600 miles; yet they say the moon performs its revolution round the earth in 27 days 7 hrs. 43 min.; but in consequence of the progressive motion (i.e. the orbital motion) of the earth, the moon takes 2 days 5 hrs. longer to again occupy the same position between the sun and the earth.

Now the earth's supposed diameter, according to the globular idea, is about 8,000 miles, and the moon's distance from the earth is stated by many leading scientists to be 240,000 miles; this distance is thirty times that of the earth's diameter. But taking the calculation that the earth is said to be eleven times and a fraction larger than the moon, and that the moon's orbit is more than 60 times the size of the earth's circumference, giving it as 25,000 miles, one cannot but stand amazed at the awful nature of the moon's supposed motion in flying round the earth; and when the two motions are in the same direction (*i.e.*,the earth's and the moon's) the velocity would consequently be tremendous! It is truly "a romance." It would be impossible for eclipses to last as long as they do on this assumption. The retardation in the one case, and the acceleration in the other case, every day, cannot be adequately described. Such motions are unthinkable, yet we are asked to accept them in place of what, to all appearance, is leisurely regular motion, similar to that of the sun, which, like fast ships, bound for America, appear to glide along—the sun in the sky, and the ships on the waves.

It is a difficult task to differentiate between the immense difference of the stated speed of the earth round the sun, and that of the moon round the earth, except to say that the proportion is as 180 to 1,100 per minute, and this differentiation goes on perpetually. It is evident that they could never be in that position to each other whereby a lunar eclipse could, at the most, last more than two minutes.

The following diagram, from *Aether and Gravitation*, by William George Hooper, F.S.S., illustrates the moon's supposed path round the earth, and also the sun's relative position to the earth and the moon. This diagram does not perfectly represent what the moon's path ought to be on the latest theories, as we have shown in a previous chapter, but it is an approximation, and we copy it as Mr. Hooper has given it.

Let us in starting represent the earth's orbit by a perfect ellipse, A B C D, with the sun occupying one of the foci, S.

DIAGRAM XIV

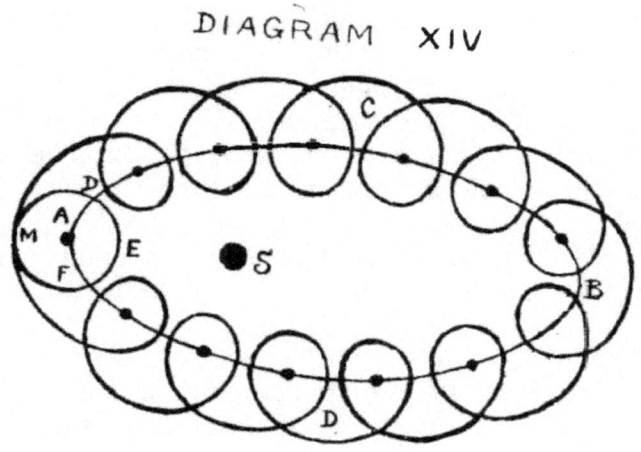

We will suppose that the earth is at point A of its orbit, and is being circled round the sun with uniform velocity. As it is circled round the sun by the sun's aetherial currents, at the same time its satellite, the moon, is being circled round the earth by the electro-magnetic Aether currents which circulate round that planet.

We will represent the orbit of the moon by part of a smaller circle, D E F, and suppose the moon to be at point D of that orbit.

As the mean distance of the moon from the earth is said to be about 240,000 miles, and the diameter of its orbit 488,000 miles; therefore the circumference of such an orbit would be 488,000 x 3'1416, which gives us about 1,533,100 miles. That distance is supposed to be traversed by the moon in about 28 days, so that the moon's average velocity in its orbit, as it is circled, or pushed round the earth is about 2,281 miles per hour.

While, therefore, the moon is travelling 2,280 miles, the earth in its journey round the sun is supposed to travel about 64,800 miles in the same time. So that by the time the moon has travelled half its orbit, that is, from D to F, which would take about 14 days, the earth has also travelled in its orbit 64,800 x 24 x 14 = 21,772,800 miles in front of that point.

In a similar way, while the moon goes on to describe the other half of her supposed orbit, the earth is still said to proceed on its journey, so that at the end of 14 days it is again 21,772,800 miles further on, with the result that the centripetal force (by which the moon is said to be attracted to the earth) keeps it at the distance of 240,000 miles according to Kepler's Second Law, as explained in Art. 103.

The moon, therefore, completes her orbit about 21,772,800 miles further on than it would do if the earth were stationary. The effect of this continual progress of the earth on the moon's orbit, as it describes its own orbit round the sun, is seen in diagram XIV.

As the moon is said to revolve round the earth thirteen times in one year, and to perform thirteen revolutions round this "planet," it cannot be said these orbits form perfect ellipses, as the earth is also said to be ever circling round its central body, the sun. Therefore this diagram, as already remarked, does not accurately represent the orbital motion of the moon through space, as it assumes that the earth returns to the same point in space from whence it started. This, however, is incorrect, as we have to remember that the sun is also said to be rushing, with the whole "solar system" with great velocity towards the star *Delta* Lyrae, as stated in Sir Robert Balls book, entitled: *In the High Heavens*, p. 26. Referring to *Delta Lyrae, the point to which the Solar System is moving*, the author says: "Every owner of a telescope is acquainted with *Beta* Cygni, the most glorious coloured double star that the northern heavens have to offer. A line from Vega to *Beta* Cygni shows, at about one fourth of the way, a bright star, which is *Delta* Lyrae. It is towards this particular spot of the heavens that the sun, bearing the earth and all the other planets with it, is hurrying at this moment." At p.31, Sir R. Ball says: "It would seem that the sun and the whole solar system are bound on a voyage to that part of the sky which is marked by the star *Delta* Lyrae. It also appears that the speed with which this motion is urged is such as to bring us every day about 700,000 miles nearer to this part of the sky. In one year the solar system accomplishes a journey of no less than 250,000,000,000 miles. As you look at *Delta* Lyrae to night you may reflect that during the last 24 hours you have travelled towards it through a distance of nearly three-quarters of a million of miles. So great are the stellar distances that a period of not less than 180,000 years would be required before our system, even moving at this impetuous speed, could traverse a distance equal to that by which we are separated from the stars." On p.32, we read: "The merest glance at the sky through a telescope will show us that our world is only one of many worlds." Therefore, according to the above, if the earth really made an annual revolution round the sun, by the time it was accomplished, the whole "solar system" would necessarily be carried millions and millions of miles through space, and consequently the earth's so-called "orbit" should *not* be represented as it is by a *perfect ellipse*.

But modern science evidently, does not lead her supporters to be very accurate in illustrating their ideas. The following diagram is one of their productions, showing the relative positions of the planets (of course including the earth !) around the sun.

# ·DIAGRAM XV

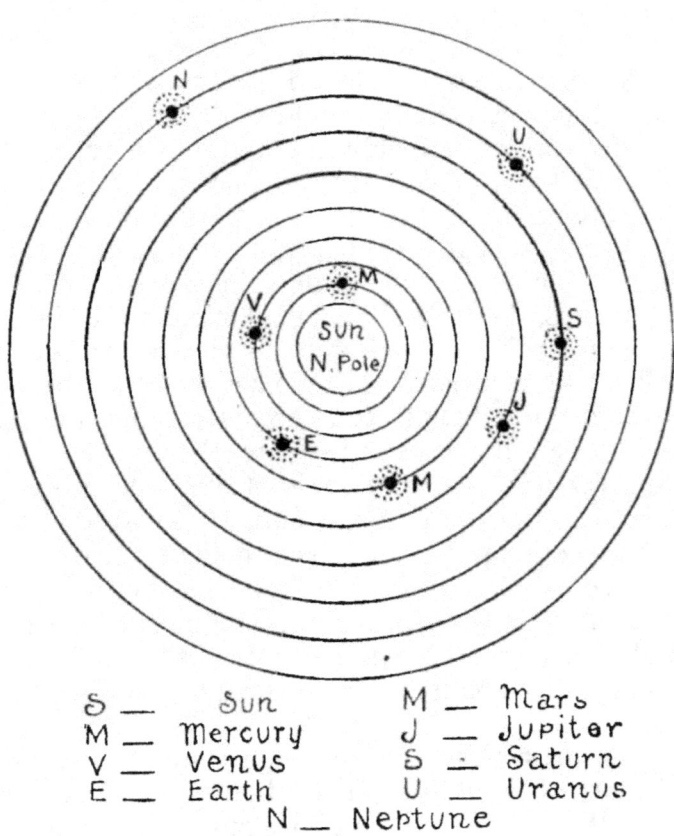

| S — | Sun | M — | Mars |
|---|---|---|---|
| M — | Mercury | J — | Jupiter |
| V — | Venus | S — | Saturn |
| E — | Earth | U — | Uranus |
| | N — Neptune | | |

This diagram is an illustration of the primary reason for astronomers adopting the theory associated with the globular hypothesis, viz.: a desire to present a mathematical reason why, for the motions of all bodies automatically, without Divine intervention. They adapted the circle to their purposes, creating a scientific formula, starting from the truism that all lines drawn from the centre of a circle to the circumference are proportionately the same.

They divided the circle into 360 parts, each part being called a Degree. A number of circles one inside the other will have the same point for their centres, *i.e.*, two lines being drawn from the innermost circle and extended, intercepting a quarter (90 deg.) of the outer circle, exactly a similar proportionate interception would apply to each of the other circles; therefore it was assumed that the bodies in the firmament are spherical (or circular) bodies, and forming circles having a common centre, *i.e.*, the sun, round which all other bodies revolve, and subjected to the two hypothetical laws of Nature: centrifugal and centripetal forces, and associated with the Laws of Gravitation and Attraction (the absurdities of which we are endeavouring to expose); hence the so-called solar system represented the planets as rings within rings, and, in order to carry out the hypothesis, the earth became included in the list of planets, the seven others being named after ancient mythological deities, and according to J. Norman Lockyer, a globe a little over two feet in diameter would represent the sun—Mercury being proportionately represented by a grain of mustard seed revolving on a circle 164 feet in diameter; Venus: a pea on a circle of 284 feet in diameter; the Earth also a pea on a circle of 430 feet; Mars: a rather large pin's head on a circle of 654 feet; Jupiter: a moderate sized orange on a circle nearly half-a-mile across; Saturn: a small orange on a circle of four-fifths of a mile; Uranus: a full sized cherry (or small plum) upon the circumference of a circle more than a mile and a half; Neptune: a good sized plum on a circle about 2½ miles in diameter.

Mercury, the nearest planet to the sun, is said to revolve round the sun at a distance of about 35 millions of miles; Venus, 66 millions of miles; the Earth, from 92 to 93 millions of miles. The mean average distance of Mars is said to be 139 millions of miles from the sun; Jupiter, 476 millions of miles; Saturn, 872 millions of miles; and Neptune, 2,746 millions of miles from the sun, and taking 60,126 days to go round the sun. How all these bodies can have one common centre passes one's comprehension—and especially seeing that now astronomers have changed the circle into the elliptical form, in respect of which a common centre is a mathematical impossibility; consequently the present phase of globular astronomy is audaciously illogical.

Summarizing these calculations, we arrive at the following conclusions:—

The moon's monthly orbit round the earth (if circular) would be about 1,533,000 miles, it follows that in the thirteen revolutions which it is said to make during the year, it travels about 19,929,000 miles, and this independently of, and over and above the distance of the earth's annual orbit round the sun, which is said to be about 588,000,000 miles. As the moon partakes of this motion also, its total distance travelled during the year must be about 607,929,000 miles. Moreover, the rate of the moon's independent motion (according to the accepted theory) would be about 2,280 miles per hour, and that of the earth being 64,800 miles per hour, it follows that the moon's velocity (if the direction be similar to a cork-screw) might be actually estimated at 67,300 miles per hour, and not 10,800 miles, as stated by the astronomers.

If that motion be in any other plane, then it would be most irregular; for when contrary to the earth's motion it would be practically standing still, and when moving in the same direction it would be immensely faster, in order to overtake and pass in front of it. In any case the actual phenomena demanded by the modern astronomical hypotheses are seen to be absolutely impossible; and when we remember that the sun is also supposed to be rushing onwards through space at the fearful rate above quoted, then the earth's orbit would no longer be direct, nor even elliptical, but would be something distinct from both, though something like a spiral movement; while the orbit of the moon would be altogether different.

These motions and velocities are unthinkable, and they write folly on the whole astronomical hypothesis. But we will now deal with

## THE MOON'S REAL MOTIONS.

Zealous endeavours have been made to find a physical basis, if not a physical cause for the idea of "universal attraction." Enough has been given by a late writer, even while making similar attempts, to warrant those who have "a competent faculty of thinking" altogether to discard so unphilosophical an idea. There can be no physical basis for the idea of the "Attraction of Gravitation," since the term necessarily implies "action at a distance."

Enlarged conceptions of the *repulsive* forces of Electricity and Magnetism, in harmony with practical experiments and experience, leave no room for the idea that bodies can pull each other at a distance without some solid and continuous pulling tackle. This pulling tackle, even if it existed, would have to be a sort of elastic, always drawing itself inwards, to answer to the idea of "attraction."

Examination of the latest astronomical theories of a moving sun proves that there cannot possibly be any *elliptical* planetary orbits with the sun as the attractor "in one of the foci." So that we Zetetics have a further reason, if one were needed, for altogether discarding the idea of "the Attraction of Gravitation," formulated to support elliptical orbits, "now" it is proved that no such orbits exist. Lest we should appear to be asserting too much we will give the following further quotation from *Aether and Gravitation*, by W. G. Hooper, F.S.S.:

"In Art. 4 we learn that Newton in the first rule states that 'Nature is simple, and does not abound in superfluous causes of things.' And again, in the nature of philosophy nothing is done in vain; and by means of many things it is done in vain when it can be done by fewer. Here, then we have apparently two forces (electricity and 'gravitation') which act in the same molecular or planetary or interstellar spaces at one and the same time. Therefore, if this be true, Nature does abound in superfluous causes, because we have two forces in existence where one will suffice, and one of them therefore exists in vain. So that it will be philosophic if we do *away* with *one of the causes*, and replace the two causes by only one. Now which shall be done away with, the electrical attraction (electrical currents really) which is due to a physical medium, the electrical magnetic Ether; or, the gravitation attraction that is caused by the virtue of some body of which *we have no knowledge*, which is transmitted through space in a way that we cannot understand, and acts upon distant bodies in a manner altogether *outside our usual experience and observation*? There can only be one answer. If either of these two forces is to be done away with, *it must be the mysterious, intangible, unphilosophic attraction of gravitation,* which must be replaced by the philosophic and known attraction (repulsion) of electricity, which can be traced to a physical medium, the Electro Magnetic/Ether." (Italics and brackets ours). *Aether and Gravitation,* p. 287.

This is excellent; and it is a pity that the writer is not consistent with himself; for, even after so pronounced and logical statement, he tries to save the face of "gravitation" by adding: "we are compelled to come to the conclusion that the attraction of Gravity and Electric Attraction are one and the same."

Th e conclusion that we come to is, that there is no need for the idea of the "Attraction of Gravitation" at all, when Mr. Hooper has so well proved that Electricity and Magnetism are sufficient to account for all the actual phenomena of the universe. Besides, he has also proved that Electricity is a *repulsive* force, and that which is repulsive is the very opposite to that which is "attractive." Repulsion is due to impact at close quarters, and harmonizes with our experience; but to what in the world is "attraction" due between bodies "at a distance"? Were it not for the globular theory Mr. Hooper's reasoning would be irresistible; but that theory, and an evident desire to save the face of the astronomical theory of Universal Gravitation considerably mars his work.

But we must say something to our readers on the moon's true motions. And, having seen the impossible nature of the moon's motions according to the latest astronomical theories, they should be all the more willing to accept those laws of motion which are in harmony with a moving sun and a motionless earth.

We have before shown that the sun, moon, and planets, all revolve daily, at comparatively short distances, around and above the earth in *spiral orbits,* approaching more or less closely to exact circles. The "fixed stars" move in exact circular orbits, whether in the north circuit, or in the south circuit, as they never vary in their declination. As the sun, moon, and planets are continually varying their declination, or distance from the celestial equator, they sometimes "spiral" around in the northern currents or vortex. Sometimes when they have south declination, they "spiral" in a similar manner round the southern vortex. *The moon's orbit therefore is that of* a double spiral.

Of course this applies to the moon's daily revolution around and above the earth, as well as her passage across the equator, due to her alteration of declination from North to South.

We have in previous chapters explained the double spiral which represents the sun's motion both in its daily revolution, and its progress through the signs of the Zodiac. The moon similarly makes a spiral daily, as her motions are due to the same causes as those of the sun and planets. We have therefore no need again to go into those points which are similar, but only into those in which the moon's motions differ from those of the sun.

The circling currents of the Aether are the cause of all planetary motion, including the motions of the sun and moon. But the planets having different sizes, and possessing different densities, occupy various positions and elevations in the Electro Magnetic currents. As these currents in various parts have different rates of motion, like the different currents of a stream; the various heavenly bodies are carried daily round the earth in different times, some in less and others in more than twenty-four hours. The sun takes twenty-four hours to circle from a given meridian to the same again. This would give to it a speed, when near the equator, of about 1,040 miles per hour. The moon takes nearly twenty-five hours to complete a daily revolution, and to return again to the same meridian; this gives her a speed, when near the equator, of about 1,000 miles per hour. Of course her actual rate varies with her declination; when her "spiral" is less her rate will be somewhat less, as is the case with all the planets.

Again, the moon being lower in the ethereal vortices, that is nearer to the earth, than the sun, her motion is actually slower, which is contrary to the astronomers' hypotheses. That she is slower may be noticed from the fact that she is about an hour later each day in rising. Then, being slower than the other heavenly bodies, the signs of the Zodiac and the constellations pass her at the rate of about 13 deg. daily, so that all pass her in about 28 days, which makes the astronomers think that the moon has gone through the signs in a direction contrary to her real motion, which contrary motion they term "direct"!

The moon therefore "crosses the equator" about once every fortnight, passing through all the signs in about one twelfth of the time it takes the sun to go through them. Thus while the sun makes nearly ninety "spirals" (almost circles) while in one of the vortices, in passing from the "Tropic" to the Equator; the moon only makes seven such spirals. This should show that the "spirals" of the moon's orbit are farther removed from the circular form, that is the spirals are coarser, or the threads of the spiral, if one may use the term, are further apart; and, whereas the sun gets to 23½ deg. both north and south of the equator, the moon seldom goes more than 18½° or 19½° declination, whether in the North or the South. This is about 5 deg. less than that of the sun; and thus the imaginary line called "the ecliptic" has been invented by astronomers, giving rise to the idea that the plane of the "Ecliptic" makes, with the Equinoctial an angle of about 5 deg.

Besides introducing the confusing term of "celestial latitude," which, as we have before pointed out, has a meaning different from that of terrestrial latitude, compared with the intricate mazes of the supposed inotions of the heavenly bodies according to the "solar system" of astronomy, the Zetetic system is simplicity itself. And, as Sir Isaac Newton remarks, in Art. 4 : "N ature is simple and does not abound in superfluous causes of things we have shown that the vortex movements of the ethereal currents, which are electric and magnetic, are quite sufficient to account for all the various phenomena of celestial motions above and around a stationary earth and sea.

As these vast currents circle round the earth, light, heat, and electricity, are but different manifestations of their one all-prevailing and ever-present force.

As the Ethereal currents circle round the two mighty vortices, which join like the figure 8, light and electricity must travel in great circles and expanding curves, and not in straight lines as commonly supposed. Indications of these great curves may be seen, not only near the earth but above in the celestial regions. Evidences of this fact may often be noticed in the direction of the moon's phases; for though the moon does not receive her light from the sun, yet her light is turned, generally speaking, towards the sun, because of the electric and magnetic currents which pass from one to the other. If watched, it is evident that these currents do not travel in straight lines, as may be seen from the following diagram .

## DIAGRAM XVI

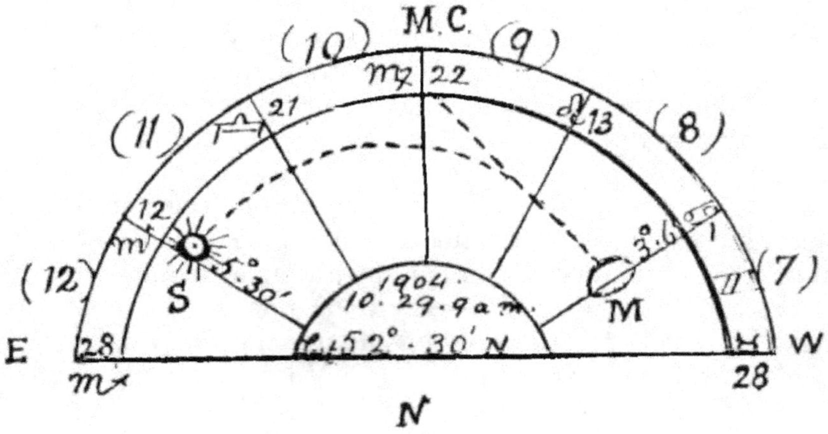

The above diagram is half the figure of the heavens, taken for and showing the respective positions of the sun and moon on October 29th, 1904, at about 9 o'clock a.m., 52 deg. 30 sec. North lat. It is a figure of that part of the heavens which was above the horizon at the time. If the whole of the heavens both above and beyond the horizon be divided into twelve parts or "houses," and those beyond the horizon northward be given the first six numbers, the rest would be as above, namely, the seventh "house," or division, would be in the

West (W), the eighth and ninth reaching up to the M.C., or mid-heaven, in the South; while the tenth eleventh, and twelve divisions would be on the eastern side of the figure (E). On this day, about 9 o'clock a.m., the sign Scorpio was rising, and the sun was near the cusp or beginning of the twelfth house in Scorpio, as seen in the figure. The moon was in the West, not far from the cusp of the eighth house, being about four houses, or 120 deg. from the sun and in 3 deg. of Cancer.

Now had the rays of the sun, or its electric currents even, travelled in straight lines to the moon, the moon's luminous face should have been directly opposed to, or facing the sun. But it was not so observed. Passing a straight line through the middle of the moon's visible face showed that she was evidently looking up towards the M.C., and not towards the sun in a line parallel to E and W. If the moon's light was merely reflected light of the sun, and his rays travelled in straight lines to the moon, the moon's phases, or face, should have been looking low, or directly towards the sun, near to the cusp of the twelfth division, in a line parallel to the surface of the earth and looking towards the east. But the moon's face was observed to be looking upwards towards the mid-heaven in the south, or meridian place of the sun at noon that same day, thus proving that neither the sun's light, nor its counterpart, the Electro-Magnetic rays proceeding from it, travel in straight lines, but in great curves, whether they come down low towards the earth, or circle high in the upper celestial regions.

After these hints, readers may watch for themselves, and make their own observations and diagrams.

Thus we have shown how all celestial motions may be explained on the basis of an electric and magnetic Aether daily circling round and above a plane and motionless earth; and, in so doing, we have brought Zetetic astronomy into harmony with our primary fact: that water is level, and the earth a plane.

The question of geography must be left over for future consideration.

We cannot conclude these articles better than by quoting the words of inspired writers: —

"The heavens declare the glory of God, and the firmament showeth His handiwork. Day unto day uttereth speech, and night unto night telleth knowledge."—*Ps.* xix. I, 2.

Again: "Thus shall ye say unto them, the gods that have not made the heavens and the earth, even they shall perish from the earth, and from under these heavens. He hath made the earth by His power. He hath established the world by His wisdom, and hath *stretched out* the heavens by His discretion..... Every man is brutish in his knowledge; (*i.e.*, in 'knowledge' based upon man's teaching in opposition to the Holy Scriptures.)"—*Jer.* x. 11, 12.

And again: "For the wrath of God is revealed from heaven against all ungodliness and unrighteousness of men, who *hold down the truth* in unrighteousness. Because that which may be known of God is manifest in them; for God hath showed it unto them. For the invisible things of Him from the Creation of the world are clearly seen, being understood by the things that are made, even His eternal power and Godhead, so that they are without excuse."—*Rom.* I. 18-20.

## "The firmament showeth His handiwork."

The outstretched heavens above appear a dome,
To everyone on earth—where'er he roam.
He sees a dome or vault above each station—
As many domes as points of observation.
Of these apparent domes there is no dearth.
Each man beholds the same above the earth.
My Zenith's highest point—just where I stand—
Forms the horizon to a distant land;
And while those far off west on sunrise feast,
My noon is someone's sunset in the east.

"And, Thou,Lord, in the beginning hast laid the foundation of the earth; And the heavens are the works of Thine hands: They shall perish; but Thou remainest: And they shall wax old as doth a garment.
And as a vesture shalt Thou fold them up, and they shall be changed: But Thou art the same."—*Heb.* I. 10-12.

A.D. 1904.

By LADY BLOUNT and ALBERT SMITH.

# THE

# EARTH

## A Monthly Magazine of Sense & Science

### Upon A Scriptural Basis,
And of Universal Interest to all Nations and
Peoples under the sun.

EDITED BY E. A. M. B.

WHICH IS TRUE?

## THE GLOBE OF LAND AND WATER

OR

## The Level Surface of Water and Land?

6 STATUTE MILES.

" Forty-three years ago, when an atheist THROUGH FALSE ASTRONOMY,
I was converted by the Spirit of God."—*Alex. McInnes.*

"The Earth" contains instructive articles, dealing with
the erroneous teaching of Modern Astronomy; proving
by the indisputable evidence of guaranteed practical
experiment on the Bedford Canal, that Water is LEVEL
and the Earth A PLANE.

Published by Lady Blount, 11, Gloucester Road, Kingston Hill.

# *Universal Zetetic Society,*

Founded in New York in Sept., 1873, and in London in Dec., 1883 (ten years after the American), as *The Zetetic Society*, by "Parallax," and others, is now firmly established by E.A.M.B., (Lady Blount), *Ed. of The Earth*, and her army of helpers, throughout the civilized world. Many local branches of the organization have been started, during the past five years, in all the principal countries, with the exception of Russia, where *The Earth* is not allowed to circulate.

PRESIDENT : LADY E. A. M. BLOUNT,

VICE-PRESIDENT : C. DE LACY EVANS,

*(M.R.C.S., Ph. D., etc., late Surgeon, Gold Coast;*
*Author of "Errors of Astronomy ;")*

(Dr. C. de Lacy Evans, was Vice-President of the Zetetic Society when first founded.)

### COMMITTEE.

| | | |
|---|---|---|
| Rev. E. W. Bullinger, D.D. | Rev. A. T. de Learsy, D D. | Albert Smith. |
| Maj.-Gen. E. Armstrong. | C. W. Makepeace, Esq | Charles Smith |
| Rev. E. W. Brookman. | Jno. S. Mc Clelland, Esq. | Isaac Smith, Esq. |
| Joseph Chamberlain, Esq. | Alex. Mc Innes, Esq. | John Smith, Esq. |
| Fredk. Evans, Esq. | Rev. E. V. Mulgrave. | H. H. Squire, Esq. |
| Elder Miles Grant. | Jonathan Nicholson, Esq. | Archbishop C. I. Stevens, |
| Dr. E. Haughton, M.D., | Dr. T. E. Reid. | D.D.,LL.D. |
| *B.A., & Sen. Moderator* | E. J. Shackleton, Esq. | A. Walter, Esq. |
| *in Nat. Sc., Trin. Col.* | A. E. Skellam, Esq. | Capt. West. |

HON. SEC. & TREAS. :

## LADY E. A. M. BLOUNT, *Ed. of "The Earth."*

11, GLOUCESTER ROAD, KINGSTON HILL, SURREY, ENGLAND ;
to whom all communications should be addressed.

### Our Motto.

## "*IN VERITATE VICTORIA.*"

### Our Object.

*The propagation of knowledge relating to Natural Cosmogony in confirmation of the Holy Scriptures, based upon practical scientific investigation.*

### RULES.

1—The so-called "sciences," and especially Modern Astronomy, to be dealt with from practical data in connection with the Divine System of Cosmogony revealed by the Creator.

2.—*Members* to subscribe Six Shillings a year, which entitles them to two copies of each issue of the Society's Organ, and a copy of every paper issued by the Society. Such will also be eligible to be voted to serve on Committees, to vote on motions, and to propose (subject to Rule 1) any alteration thought to be beneficial to the Society.

3.—*If any lover of Truth desires to become a member of the Universal Zetetic Society, and cannot make it convenient to pay a subscription, it need not deter him or her from joining. Your help will be appreciated in any way that you can give it. Each one can at least help in making known the truth.*

4.—All subscriptions to the Society to be made to the Treasurer, addressed to "Zeteo," at the office of *The Earth.*

Copies of "The Earth" (the Society's Organ), may be had of the Ed., E.A.M.B. (Lady Blount), 11, Gloucester Road Kingston Hill, Surrey, England.

www.ingramcontent.com/pod-product-compliance
Lightning Source LLC
Chambersburg PA
CBHW070109210526
45170CB00013B/806